The Sustainability of Renewable Energy in Europe

Simona Bigerna • Carlo Andrea Bollino
Silvia Micheli

The Sustainability of Renewable Energy in Europe

Springer

Simona Bigerna
Department of Economics
University of Perugia
Perugia
Italy

Carlo Andrea Bollino
Department of Economics
University of Perugia
Perugia
Italy

Silvia Micheli
Department of Economics and Business Science
Guglielmo Marconi University
Rome
Italy

ISBN 978-3-319-12342-4 ISBN 978-3-319-12343-1 (eBook)
DOI 10.1007/978-3-319-12343-1

Library of Congress Control Number: 2015932871

Springer Cham Heidelberg New York Dordrecht London

Printed on acid-free paper

Springer is part of Springer Science+Business Media (www.springer.com)

Preface

The main guidelines for energy and environmental policy have been the same for a long time now. The development of the energy sector is essential for economic development and competitiveness of all countries in the world and this development must be sustainable in terms of natural resources depletion and environmental quality.

This perspective is an ongoing challenge not only to the basic paradigm of efficient market decentralization and infinite supply of inputs in the long run, but also to the imperfect market and externality models. There is now recognition that the very survival of our society is at stake if we do not solve the issue of sustainability of our economic development. Indeed, the acknowledgement that fossil fuel reserves are finite or that emissions are dangerous to our health is not enough. We need comprehensive policies, credible policies, which can spur the necessary investment for the future of the Earth.

But if the objectives are clear—renewable energy sources development, competitiveness, diversification, energy saving—we must be humble enough to understand and study the complex relations between targets and instruments, in order to be able to offer instruments that are effective, that cost less, and that are adapted to the evolution of the economic system.

This book offers a very stimulating analysis of the most relevant issues of future European development of renewable energy sources, analyzing the available policy options.

In the first two chapters, the authors review the European policy strategies, assessing how much we have accomplished and what still needs new thinking and action.

In the next three chapters, the discussion of European policies for renewable energy sources is cast clearly to study the quantitative burden needed to support the instruments. In this respect, a crucial question seems to emerge from this book. We know that there is need for new investment in both physical capital (hardware and finance) and human capital (R&D and inventions), but will we be able to deal with the sacrifices necessary to make sustainable the development of our society?

I leave it to the reader to discover in the last chapter the answers to this question, namely when the authors discuss new scenarios for renewable energy sources development and advance precise policy proposals, which sound ethically appealing, but politically provocative.

In fact, they call for coherent and cooperative policies among all EU member countries, which can minimize the overall cost for the EU, i.e., for the entire European society. However, this requires overcoming national egoism and adopting a collective vision for the future.

Pier Carlo Padoan

Contents

Chapter 1
The EU Legal Frameworks

Energy and environment today are the subjects of a number of initiatives and programs by the European Union (EU). Climate change has become the main pillar of EU policies aimed at accelerating the transition towards sustainable development, a future with low emission of greenhouse gases (GHG). The development of multidisciplinary and multidimensional strategies by the community institutions to face climate change has occurred and is still evolving within a particularly complex system. This chapter provides an analysis on current EU policies, with targets up to 2050, and prior agreements that led to the current ones. The analysis starts from the recent agreement called "roadmap for moving to a competitive low-carbon economy in 2050" promoted by the EU, and focus on the 2030 framework for climate and energy policies presented on January 2014 by the European Commission (EC 2014) as an intermediate step to achieve the goal of longer term to 2050. The chapter then proceeds backwards, illustrating the main agreements that have led the EU to the current situation, that are the 2020 climate and energy package and the implementation of the Kyoto Protocol at EU level. Finally, it provides a description of trends in carbon dioxide (CO_2) emissions at global and European levels.

1.1 Road Map for Moving to a Competitive Low-Carbon Economy in 2050

The evolution of the energy and environmental policy of the EU is an ongoing process for the changes related to energy and environmental markets. The European Commission has recently presented as part of the Europe 2020 strategy, the flagship initiative "A resource-efficient Europe" (European Commission 2011a); it proposes a series of long-term strategic plans in areas such as transport, energy, and climate change. The Europe 2020 strategy for smart, sustainable, and inclusive growth includes five main objectives that define the goals that the EU should achieve by 2020. One of these concerns the climate and energy: the member states have committed themselves to reducing GHG emissions by 20 %, to reach a 20 % share of

S. Bigerna et al., *The Sustainability of Renewable Energy in Europe*,
DOI 10.1007/978-3-319-12343-1_1

renewable energy sources (RES) in the EU mix energy, and to get a 20% increase in energy efficiency by 2020. Currently, the EU has already achieved the first two objectives, and there are all the conditions to achieve an improvement in energy efficiency by 2020. The worldwide objective related to climate change is to contain within 2 °C global warming produced by climate change. The 2 °C cap on global warming was established in the Copenhagen Conference on climate change in 2009. The agreement has replaced by the Kyoto Protocol which would have been expired in 2012. The ambition of the conference was to be able to extend the Kyoto targets in 2020 or 2050, engaging in a binding even those nations which had not acceded to the Kyoto Protocol (i.e., the USA) or who were exempted from the cuts in emissions of GHG for not curbing their growth (such as China, India, Brazil, emerging nations among the developing countries). In this framework, the European Commission reconfirmed in February 2011 the EU objective of reducing GHG emissions by 80% by 2050 compared to 1990 levels.

To this end, the European Commission on 22 January 2014 has recently proposed new and important goals for 2030 through the "2030 framework for climate and energy policies" as an intermediate step to achieve the goal of longer term to 2050 (European Commission 2014).

The key elements of the climate and energy package in 2030 established by the European Commission are as follows.

- *A binding goal of reducing GHG.* Essential element of EU policy, it establishes a 40% reduction in GHG emissions compared to 1990 levels, to be achieved through action at national level. The reduction of the maximum annual emissions of the sectors included in the EU emissions trading system (ETS) would increase from the current 1.74 to 2.2% after 2020. The emissions from sectors not covered by the ETS should be reduced by 30% below the level of 2005, with efforts equitably shared among the member countries.
- *A binding target on the overall consumption of RES.* On the basis of a more market-oriented approach, the EU abandons national targets for an overall EU target of 27% of RES by 2030. To member states will thus be accorded some flexibility on how to transform their energy system. This is certainly an important and controversial point since the new directive does not contain any mandatory target for individual nations in terms of RES, but only a target "to achieve all together."
- *Energy efficiency.* The role of energy efficiency in the context in 2030 will be further considered with a review of the Energy Efficiency Directive ready by the end of 2014. The European Commission will assess the potential need for changes to the existing rule once the review is completed.
- *New indicators.* The European Commission proposes a new set of key indicators to assess progress over time, such as the energy price gap between the EU and major trading partners, diversification of supply and self-supply of domestic energy sources, and the ability of member states' liaison.
- *Reform of the EU ETS.* The European Commission proposes to establish a reserve for the stability of the market at the beginning of the next trading period of the EU ETS in 2021, in order to deal with the surplus of allowances that is made over recent years, to improve the system's resilience to shocks, by automatically adjusting the supply of allowances to be auctioned.

- *Governance*. The European Commission proposes new system of governance based on national energy plans proposed by the member states and coordinated by the Commission.

The 2030 framework for climate and energy policies package is raising some controversy among member states. The first observation is on clarity of objectives: only two of them, reducing the GHG emissions by 40% and expanding the share of RES up to 27% compared to 1990, are expressed in quantitative terms. The other objectives are not yet associated with measurable commitments and therefore, at this stage, appear more generic than the first two.

In addition, only the first target concerning GHG emission reduction is binding both for the EU as a whole and for member states, while the second objective relating to the RES is binding only for the EU but not for member states.

Can a macro entity, which is the sum of *n* micro entities, hit a target if its constituents are not subject to the same constraint? Many environmental problems share the features that when many countries are involved in international agreements, it is possible that some of them do not actually participate (Cremer and Gahvari 2002; Kolstad 2000; Stavins 1995). Participation to environmental agreements is related to the gains countries can get: climate change is a global problem and each country believes to get only small benefits compared to its effort.

With the 2030 framework for climate and energy policies package, there might be problems regarding cooperation to achieve RES' goal: when member states differ from each other with respect to costs and benefits of implementing RES, some of them might find affordable to free ride.

Moreover, the characteristic of public good of environmental policy implies that free riders may benefit from others' efforts without bearing any costs.

Most of the literature on free-riding states that if all countries are convinced to cooperate, an efficient treaty would specify the joint abatement policy corresponding to the internationally optimal objectives (actually, this is what happened previously with the binding targets per member state within the 2020 climate and energy package. It follows a description in Sect. 1.2). For some countries, however, the environmental policy may be so costly that it makes them worse off to cooperate, compared with the noncooperative equilibrium, i.e. the business as usual (BAU) scenario.

Another controversial aspect concerning the not-binding RES' goal at national level of the 2030 framework for climate and energy policies package is related to RES production decision. On the one hand, the lack of national constraints allows to choose the best RES to be implemented throughout the EU taking into account their productivity compared to the territory, that is geo-referencing at regional-level RES' production. For instance, each EU member states is characterized by different levels of solar radiation, wind intensity, and crop productivity for biomasses which reflects in different productions' marginal costs.

On the other hand, multinational companies in the electricity market can leverage their oligopolistic position and locate RES production in a suboptimal way. It is tricky to say whether or not liberalized electricity markets are likely to adopt the cost-saving, GHG-reducing technologies. If an investment opportunity exists to increase profit for a private power generating company, it is likely that the company will adopt it. RES' systems are in general characterized by high capital costs, low operation and maintenance costs, and zero fuel costs with the exception of systems using purchased biomass, but they are still more expensive than conventional sources. Then, the package should be able to induce participation and compliance, and, without binding objectives per member state, it might be hard to achieve the goal.

The EU had already embarked on the path of the indicative and nonbinding expansion of RES with the 2001/77/EC Directive (European Commission 2001; European Parliament 2001) on the promotion of electricity produced from RES in the internal electricity market with the indicative target of 12 % of European gross domestic energy consumption from RES in 2010, and the result was not encouraging because the target was not reached. On the other hand, at world level, something similar happened with the indicative target of Kyoto Protocol (a detailed description of the Kyoto Protocol is provided in Appendix 1).

Another observation that emerges from the member countries is the consistency of the objectives in relation to the overall EU strategies. By looking at the long term, the road map calls for an 80 % reduction in GHG emissions compared to 1990 levels by 2050. More complex is the assessment of the quantitative consistency since the stages of the path to the target of 80 % GHG emission reductions are as follows. Kyoto: 8 % in 2010; climate and energy package: 20 % in 2020; 2030 framework for climate and energy policies: 40 % in 2030; target 2050: 80 %. The succession of goals highlights two periods: until 2020, it requires a reduction of 10 % every decade; after 2020, the reduction required for each decade becomes 20 %.

Beyond question, however, very important on the feasibility of the final objective (80 % GHG reduction by the year 2050) and the definition of consequential intermediate targets, it can be said that the main target of the 2030 framework for climate and energy policies, i.e., the reduction of emissions by 40 %, is nothing more than a consequence of the long term. On this level, the EU demonstrates consistency: the target for 2030 is not rhetoric but a result of a commitment already declared.

Then, looking beyond 2020, existing and currently planned measures could bring the EU onto a pathway to achieving long-term emission reduction objectives. Anyway, achieving a reduction of emissions by 80 % by 2050 compared to 1990, as agreed by European Commission, will require enhanced efforts from member states.

The road map to achieve the objectives to 2050 indicates a gradual transition and effective in terms of cost, as indicated by the COM 112 (European Commision 2011b) on Roadmap for moving to a competitive low-carbon economy in 2050.

The COM 112 (European Commission 2011b) with the White Paper on Transport and the Energy Efficiency Plan is one of the key documents on the efficient use of resources, and it illustrates the cornerstones on which actions promoted by the EU should be based to facilitate the transition to a competitive low-carbon economy by 2050. The road map sets out a series of steps aimed at verifying whether the EU is on track to achieve its objective, policy challenges, the need for investment, and

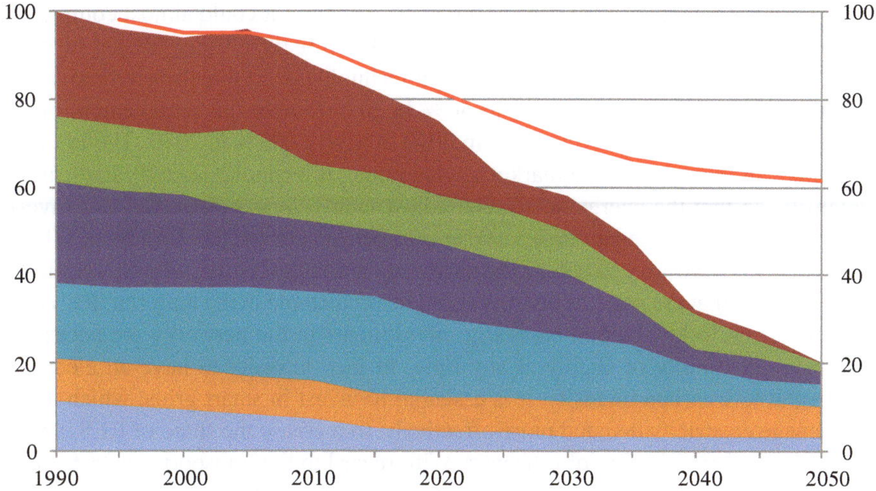

Fig. 1.1 GHG emissions (%) of the EU towards an 80% in internal reduction (100%=1990). Non-CO_2 GHG. Although the majority of GHG emissions comes from CO_2, non-CO_2 GHG such as methane, nitrous oxide, and fluorinated greenhouse gases contribute significantly to climate change. *GHG* greenhouse gases, *EU* European Union. (Source: COM 2011b, p. 112)

opportunities in the various sectors. To make the transition to a competitive low-carbon economy, the EU must be prepared to reduce its domestic GHG by 80% by 2050 compared to 1990 levels.

The European Commission has suggested, through an accurate analysis of modeling, possible scenarios of intervention to achieve this objective. One effective option would be to implement effective internal GHG reductions by 25% by 2020, 40% by 2030, and 60% in 2040. With this approach, compared to 1990, there would be an annual reduction of about 1% in the first decade (until 2020) of 1.5% in the second decade (2020–2030), and 2% in the last 2 decades until 2050.

The GHG emissions reduction would increase over time due to the increase in technology with a good cost-effectiveness relation. The path that would lead to an 80% reduction by the year 2050 is shown in Fig. 1.1.

The projection "BAU" (business as usual) in the upper part of the graph shows what would be the trend of the EU GHG emissions with only existing policies and without the implementation of the road map to 2050. The road map scenario consistent with a reduction of 80% domestic shows what would be the trend of global and sectoral GHG emissions whether additional strategies are adopted on the basis of technological options available in 2050.

The European Commission's analysis has also explored possible ways for action for the most important sectors, examining a number of scenarios based on different rates of technological innovation and fossil fuel prices. They have arisen convergent results regarding the extent of the reductions needed in each sector in 2030 and 2050. In particular, the COM (2011b) 112 states the electricity sector will play a central role in the low-carbon-intensity economy; it could almost completely eliminate CO_2 emissions by 2050 and replace fossil fuels in transport and heating. It is estimated that the share of low-carbon technologies in the electricity mix will increase from about 45% in 2011 to 60% in 2020 (due to the achievement of the target for RES) and from 75 to 80% in 2030 to nearly 100% in 2050. The EU ETS will play an essential role to market a wide range of technologies with low carbon intensity, so that the energy sector can adjust their business strategies and investment to evolution of technologies and energy prices. For the EU ETS to fulfill this function, it is necessary to reduce the maximum amount of emissions allowed under the ETS, and that the price of permits could be predictable in the long run. RES play a crucial role in low-carbon economy: investments in the networks are necessary to ensure continuity of supply at any time. In fact, in order to have an electrical system at low carbon intensity, it is essential to invest in smart grids, which allow to manage electricity demand more effectively, to increase the share of RES, and to achieve the electrification of transport. With regard to the transport sector, technological innovation can lead to an efficient and sustainable transport system through the introduction of biofuels and the efficiency of the vehicles. The built environment offers the possibility to reduce emissions in the short term, primarily due to improved energy efficiency in buildings. The analysis conducted by the European Commission shows that in this area it would be possible to break down GHG emissions by approximately 90% by 2050, which represents an above-average contribution in the long term.

Concerning the industrial sector, its GHG emissions could be reduced by 83% in 2050. The use of plants and industrial processes more efficiently in terms of energy and resources, an increased recycling, and the use of technologies for the reduction of non-CO_2 emissions could provide a decisive contribution in polluting emissions reduction.

Even the agricultural sector will be able to cut its non-CO_2 emissions by between 42 and 49% compared to 1990 levels by 2050, thanks to the efficient use of fertilizers, the biogasification and a better management of organic fertilizers, diversification, and marketing of production at local level.

Different energy sources with low carbon emissions with associated systems and supporting infrastructure, including smart grids, carbon capture and storage, advanced industrial processes, and the electrification of transport (including energy storage technologies), are the pillars of energy policy for the period after 2020.

1.2 The 2020 Climate and Energy Package

In order to understand the 2030 framework for climate and energy policy, it is important to analyze and understand the 20–20–20 climate and energy package. In early 2007, the EU proposed an integrated energy and climate change package to cut emissions for the twenty-first century, based on a strong commitment in favor of a low energy consumption economy, more competitive economy, and sustainable development. The novelty of the approach lies in the centrality of the issue of the integration of energy policy with environmental and with the Lisbon strategy[1] for competitiveness. On 23 April 2009, it officially adopted a legislative package called 20–20–20 climate and energy package, through which the EU is proving to be ready to assume the global leadership to tackle climate change, to address the challenges of energy security, making Europe a model of sustainable development for the twenty-first century.

An important merit of the package is the introduction of legally binding targets to be achieved by 2020:

- A reduction in GHG emissions by at least 20 % compared to 1990 levels, with the aim of further reducing these emissions by 30 % by 2020 if other developed countries commit themselves to comparable efforts.
- Raising the share of EU energy consumption produced from RES to 20 %.
- A 20 % improvement in the EU's energy efficiency.

The targets were set by EU leaders in March 2007, when they committed the EU to become a highly energy-efficient, low-carbon economy, and were enacted through the package in 2009 (Directive 2009/28/EC). Targets are calculated with reference to 2005 taken as the base year with respect to which they are presented increases or reductions of shares of energy from RES and GHG emissions. The package responds to the appeal launched by the European Commission on real actions for the transition to sustainable development.

The package includes a series of closely related major policy proposals, such as:

- The revision of the directive on the EU ETS.
- A proposal on the allocation of efforts by member states in order to reduce GHG emissions in sectors not covered by the EU ETS (such as transport, buildings, services, smaller industrial installations, agriculture, and food industry).
- A directive on the promotion of RES to achieve the targets for reducing GHG emissions.

The importance of the EU ETS lies in creating a market and an artificial price for a public good like clean air. Firms subject to the EU ETS have to face costs when they emit CO_2 emissions: on the one hand, if a firm needs for its activities a larger number of permits than it has available, then it must incur a cost to buy them; on the other hand, opportunity costs increase because permits can be sold in case of nonproduction.

The changes made by the package on the system of emissions trading shall come into force during the period 2013–2020, and they are done in order to overcome the problems arisen during the early years of implementation within the Kyoto Protocol.

[1] The European Council on 23 and 24 March 2000 in Lisbon has launched a series of ambitious reforms that have taken the name of the "Lisbon Strategy" for the years 2000–2010, in order to strengthen employment, economic reform, and social cohesion in the context of knowledge-based economy. The priority areas identified by the Lisbon Strategy are investments in knowledge and innovation, unlocking business potential, especially of small and medium enterprises, employment of priority categories, and energy policy for Europe.

The first problem is related to the European mechanism of allocation used so far. Emission allowances are allocated free of charge, and the allocation could be made on the basis of past emissions (called grandfathering). This mechanism can create a vicious circle because it does not stimulate the adoption of new technologies with low environmental impact. Furthermore, it favors large companies that in the first phase of allocation received a large number of permissions to maintain their level of activity at the expense of small companies (Borghesi 2010).

Another problem is related to inconsistencies between emission permits and national allocation plans (NAP): governments have created too many emissions permits to protect the welfare of domestic firms who wish to receive the highest number of permits.

Finally, the large and persistent fluctuations in market prices have caused havoc in the market and uncertainty about the quality of environmental policy.

The main elements of the new system of EU ETS are:

- New emissions cap set at 20 % less than 2005 levels by 2020.
- The use of credits from the clean development mechanism (CDM) and joint implementation (JI) is limited to 50 % compared to the overall reduction in emissions of the EU in the period 2008–2020.
- The inclusion of new sectors such as aviation and the aluminum industry.
- Firms operating in the electricity sector are obliged to buy 88 % of the allowances to each plant through an auction mechanism; 10 % of the permits is redistributed from countries with higher per capita income to low per capita income countries and the remaining 2 % is given to the member states that have failed to achieve the target of 20 % GHG reduction in 2005 (i.e., the Eastern European countries).

The sectors covered by the EU ETS are thermal power plants, industries in the field of energy production and manufacturing production (energy activities, production and processing of metals, cement, ceramics and tiles, glass, paper), and aircraft operators. From 2013, the sectors have been further extended to the production of aluminum, lime, nitric acid, hydrogen carbonate, and sodium bicarbonate and facilities that deal with the capture, transport, and storage of CO_2.

The adoption of the auction mechanism in the ETS means a better distributional effect compared to grandfathering, since the revenue generated from the auction can be used either to reduce distortionary taxes or to promote research and development in clean technologies (Carraro et al. 2006).

The most important point concerns the directive on RES to achieve the goal of 20 % on energy consumption by 2020 and to share the burden among member states. The RES Directive has innovated the Kyoto Protocol which left to the discretion of the countries the choice of methods to achieve the nonbinding objectives, while the directive sets the target of 20 % by 2020 divided into subnational binding targets for member states (Fig. 1.2). In particular, 50 % of this effort must be shared equally among the member states, while the remaining 50 % is modulated according to gross domestic product (GDP) per capita. The objectives are then modified to

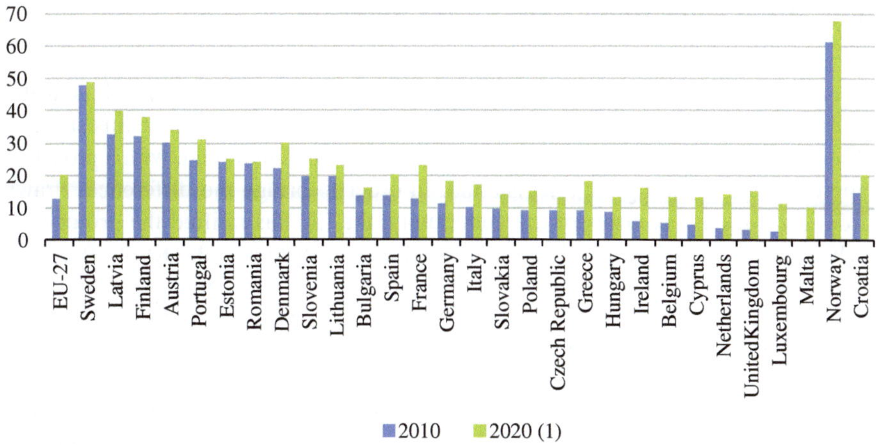

Fig. 1.2 Share of RES (%) in EU gross final energy consumption, 2010 and 2020. (1) Legally binding targets for 2020. *RES* renewable energy sources, *EU* European Union. (Source: Eurostat 2014)

take into account efforts already made by the member states which have increased the share of fuels from RES in recent years.

With regard to electricity, the European Commission estimates that RES might rise from 15 % in 2009 to 35 % in 2020. The promotion of EU electricity production from RES takes place in a market that is increasingly competitive. Based on the experiences of liberalization of the electricity market in the world, the EU's objective is to achieve greater efficiency and lower consumer prices through the introduction of competition, although it is rather difficult for firms to produce energy from renewable sources to compete in the energy market based primarily on energy from fossil fuels.

The liberalization of the electricity market may appear as a partial response to climate change as it allows consumers to purchase cleaner electricity directly from suppliers; however, the majority of consumers are not willing to pay higher prices for electricity from renewable sources as they are burdened by higher prices in order to preserve a public good (i.e., clean air), all of which benefit (Bigerna and Polinori 2014).

Overall, the efforts of member countries in achieving the objectives of the package have already led to important results. The situation has now changed radically since the end of 2008: the reduction of GHG and the RES' objective among the package are the area in which the goals have already almost been achieved in 2013; things are less well with regard to the 20 % improvement in energy efficiency and the EU is much below the target. Even with the third goal away from reaching in 2020, the European Commission has adopted the new proposals by 2030 (compared to 1990).

1.3 EU Legislation on the Kyoto Protocol

In order to understand the success of 2020 climate and energy package, it is useful to deepen the basis, further back in time, that have allowed to achieve this success. The EU has always been very proactive in the environmental field, with a proactive approach and a cutting-edge legislation. In order to bridge the differences drawn from the various countries of the community, the EU has ratified the protocol itself collectively on behalf of all member states. According to the Kyoto Protocol, industrialized countries and countries with economies in transition need to reduce their total emissions between 2008 and 2012 at least 5% compared to 1990 levels. World CO_2 emissions from fuel combustion before the entry into force of the Kyoto Protocol and its targets are shown in Table 1.1.

The EU has become a pioneer of the EU ETS in 2005, three years before the date set by the United Nations Framework Convention on Climate Change (UNFCCC) for the start of the international trade in shares. The EU has the advantage of having good mechanisms of governance and control in order to ensure compliance with the common rules by the member states. The EU has several regulatory tools to control and determine the various policies. These instruments, called EU law, are based on the treaties establishing the European Communities. The EU law may be adopted by the council of the EU or jointly by the European Parliament, and it can take many forms. In particular, "regulations" are binding legislative instruments that are applied directly within the member states, without the need for a national transposition or other steps by states. Regulations are called "European laws." The "Directives" are intended for one or more member states and must be transposed in national legislation in accordance with the procedures of the various states. The "Decisions" are legislative acts addressed to one or all member states. There are also the "Recommendations" and "Communication" which are non-binding acts of the EU, and also the "messages" that have a purely informational.

Community legislation on the Kyoto Protocol is already equipped with a system of rules and procedures for verification and compliance. The most widely used legislative instrument by the EU on the Kyoto Protocol is the directive.

The following are the most important texts of the EU legislation around the Kyoto Protocol.

Directive 96/61/EC on integrated pollution prevention and control sets out measures to prevent or reduce emissions that affect pollution of air, water, and soil. The main point of the directive is the configuration of a system of permits issued by environmental authorities in each member state for authorization of emissions by large manufacturing and combustion plants (European Commission 1966).

Decision 1999/296/EC establishes a mechanism for the monitoring of emissions of CO_2 and other GHG in the community and to evaluate the progress made towards the fulfillment of commitments to reduce emissions. The Decision 1999/296/EC also sets guidelines for the annual reports of the member states to the European Commission, and it requires states to plan and implement national programs for reductions in GHG emissions (European Commission 1999).

Table 1.1 World CO_2 emissions from fuel combustion (Mt) and Kyoto targets. (Source: IEA 2010)

	1990	2008	% change 1990–2008	Kyoto target (%)
Kyoto parties with targets	8858.3	7980.1	−9.2	−4.7%
North America	432.3	550.9	27.4	
Canada	432.3	550.9	27.4	−6
Europe	3153.6	3222.9	2.2	
Austria	56.5	69.3	22.7	−13
Belgium	107.9	111.0	2.8	−7.5
Denmark	50.4	48.4	−4.0	−21
Finland	54.4	56.6	4.0	0
France	352.3	368.2	4.5	0
Germany	950.4	803.9	−15.4	−21
Greece	70.1	93.4	33.2	+25
Iceland	1.9	2.2	17.0	+10
Ireland	29.8	43.8	46.7	+13
Italy	397.4	430.1	8.2	−6.5
Luxembourg	10.5	10.4	−0.6	−28
The Netherlands	155.8	177.9	14.1	−6
Norway	28.3	37.6	33.0	+1
Portugal	39.3	52.4	33.5	+27
Spain	205.8	317.6	54.3	+15
Sweden	52.8	45.9	−13.0	+4
Switzerland	40.7	43.7	7.4	−8
UK	549.3	510.6	−7.0	−12.5
Pacific	1346.6	1582.0	17.5	

Table 1.1 (continued)

	1990	2008	% change 1990–2008	Kyoto target (%)
Kyoto parties with targets	8858.3	7980.1	−9.2	−4.7%
Australia	260.1	397.5	52.9	+8
Japan	1064.4	1151.1	8.2	−6
New Zealand	22.0	33.3	51.5	0
Economic in transition	3852.9	2624.3	−31.9	−6
Other countries (nonparticipating)	**11,566.6**	**20,368.2**	**76.1**	
USA	4868.7	5595.9	14.9	−7
China	2244.4	6550.5	191.9	
Latin America	869.5	1476.5	69.8	None
Africa	545.6	889.9	63.1	None
Middle East	592.5	1492.3	151.8	None
Non-OECD Europe	106.1	92.2	−13.1	None
Asia (excl. China)	1510.1	3521.1	133.4	None
World	*20,964.8*	*29,381.4*	*40.1*	

OECD The Organisation for Economic Co-operation and Development

Decision No. 1600/2002/EC of the European Parliament and the council lays down the Sixth Environment Action Program for the period 2002–2012 (European Commission 2002b). Such decision, defining climate change as a top priority for community action, requires a demonstrable progress in fulfilling the commitments set by the Kyoto Protocol.

Decision 2002/358/EC, also known as burden-sharing decision, has approved the Kyoto Protocol on behalf of the entire community with the symbolic deposit of instruments of ratification simultaneously by the member states on 1 June 2002. The most important measure contained in Decision 2002/358/EC is redistribution of individual targets for reduction of GHG emissions between the EU-15, the so-called burden sharing (European Commission 2002a).

The heart of the EU legislation on the development of the Kyoto Protocol is the Directive 2003/87/EC (European Commission 2003) that has introduced a cap and trade market. The directive has been formulated for the establishment of a system for the exchange of GHG emission quotas, called EU ETS, and it also defines a framework for national policies aimed at reducing emissions. The goal of the 2003/87/EC Directive is to give member states a competitive advantage, starting to apply the provisions of the protocol relating to the emissions trading 3 years earlier than the date set by the UNFCCC (2005 instead of 2008). The directive refers to any matter relating to the generation of a structure permits or allowances and a market for such permits to industrial plants that emit large amounts of CO_2.

The Directive defines in particular:

- The constitution from 1 January 2005 of a system of permits for the right to emit CO_2 for firms engaged in activities listed in the Directive. Permits indicate mode and frequency requirements for monitoring of emissions, and they are bound by the obligation to return a number of allowances equal to the emissions released by plants every year.
- The draft of NAP by all member states. The NAP prepared initially for the period 2005–2007 will subsequently be drawn up every 5 years and indicate how each country intends to reduce its GHG emissions.
- Granting and validity of emission allowances.
- Transfer and cancelation of allowances.
- Guidelines for the monitoring, reporting, and verification of emission levels.
- Penalties to be applied to noncompliant firms. In the period 2008–2012, the sanctions amounted to € 100/tCO_2. In addition, the payment of the fine does not exempt the firm from the obligation to surrender allowances due, which must be paid within the next year.
- The designation of a competent authority.
- The establishment of a system of records for the issuance and control of emissions allowances.
- Reports by member states.

From 1 January 2005, all industrial operators and companies have the power to emit CO_2 only with a permit granted by the national authority.

The phases of the ETS market, provided by the EU Directive 2003/87/EC, are the following:

- Phase I: pilot period which anticipated the date of entry into force of the obligations under the Kyoto Protocol, with a duration 2005–2007.
- Phase II: 2008–2012 duration, coinciding with the period of commitment of the Kyoto Protocol.

The EU ETS represents the main element of the European Climate Change Program launched by the EU in order to reach the Kyoto Protocol goals.

The international trade's emissions includes the buying and selling of credits among developing and/or industrial countries to fulfill its obligations under the Kyoto Protocol, which aims to reduce emissions through the commercialization of carbon credits between countries in surplus and those in deficit. The EU ETS aims to help EU member states to fulfill their obligations to limit or reduce emissions of GHG in a cost-effective manner. The fact that the firms participating in the scheme can buy or sell emission rights gives the possibility to reduce emissions at a lower cost compared to the mechanisms.

With the introduction of the cap and trade mechanism of CO_2 emissions, there have risen different platforms on which it is possible to trade emission rights. The market for emission rights is kind of over-the-counter (OTC) market, that is, it does not apply obligation to concentrate trading in a single market regulated and supervised by a public authority; the creation and management of a market is left to the business venture. Despite this, the greatest concentration of trade took place on the UK market ECX (European Climate Exchange) in which, for instance in 2009, they focused as much as 88 % of the total volume traded.

On the ECX market, there are two types of stocks that are traded: the European Union Allowances (EUAs) and certified emission reduction (CER). Both stocks give the right to emit 1 t of CO_2, but differ in their origin and their usage limits:

- EUAs are rights to emit CO_2 resulting from the allowances allocated to individual firms by national governments with the preparation of the NAP of each phase.
- CER are rights whose origin is external to the quotas allocated by governments with the NAP; these credits are earned in CDM projects.

Emission allocation mechanism involves at around 50 % of the total CO_2 emissions in Europe; within the EU ETS, electricity market plays an important role. There is no consensus among scholars on the economic impact of the ETS scheme on industries related to such allocation, especially on the effects of the emission trade system both on the performance and strategies of the firms involved. Grubb and Neuhoff (2006) show that the CO_2 allocation scheme might generate distortion mechanisms in users decisions; however, electricity market seems to be the only with a positive balance due, on the one hand, to the market power (electricity market has low competition at international level). On the other hand, CO_2 allowances might reduce generation costs since users who have a choice may be willing to diversify their portfolio energy mix. The empirical literature is inconclusive on the extent of ETS prices on electricity prices.

It also deserves to be mentioned briefly about Directive 2004/101/EC the so-called "Linking Directive" (European Commission 2004). This is an amendment to Directive 2003/87/EC aimed to enable the exchange within the EU of credits from JI and CDM projects (European Commission 2003). The purpose of the directive is to increase the trade of ETS within the EU and to reduce the total costs of implementing the Kyoto Protocol.

The important points of the Linking Directive are the following.

There is a limit to the use of CERs/emission reduction units (ERUs) within the EU ETS. Each member state shall decide on the use of such instruments in accordance with the supplementarity principle indicated by the convention, according to which the use of flexible mechanisms must be supplemental to domestic actions to reduce GHG emissions. Moreover, since 2008, the national governments determine a limit on CERs and ERUs converted by firms and show it in the NAPs. The supplementarity principle is subject to the supervision of the European Commission, on which member states are obliged to report on national actions undertaken in terms of GHG reductions. The NAP must contain both the expected amount of CERs and ERUs that can be used by operators and the percentage of allocation to each firms within which managers are allowed the use of ERUs and CERs in the EU scheme during the period considered. CERs and ERUs generated from nuclear power plants are not recognized valid for the purposes of the exchange of emission credits within the EU market, as well as reductions in emissions from carbon uptake activities of land use, change in its land use, and forestry. CERs and ERUs from projects for the production of hydroelectric power with a generating capacity exceeding 20 MW shall be considered valid if these projects meet certain international rules on the protection of dams, or the requirements established by the World Commission on Dams. There is no double counting of units: member states may not issue CERs and ERUs from project activities, which in turn lead to reductions or limitations of direct or indirect GHG emissions achieved through projects carried out by the installations covered by the ETS Directive, unless an equal number of allowances is deleted from the national register by the firm considered.

The economic rationale at the bottom of the flexibility mechanisms in general lies in the exploitation of differences in marginal costs of interventions aimed at reducing GHG emissions in the different parts of the framework convention.

The Kyoto Protocol establishes mechanisms but it does not contain specific rules on them and how they work. For example, it sets no upper limit to the size of the total contribution of GHG reduction that may be satisfied through the use of flexibility mechanisms rather than through domestic initiatives and actions to reduce GHG emissions. Nevertheless, the Kyoto Protocol explicitly provides for the requirement of supplementarity, according to which the parties should give priority to domestic actions that represent a significant part of their overall efforts in the fight against climate change.

The Kyoto Protocol, therefore, proposes a set of tools for action: strengthening of national policies to reduce emissions by improving energy efficiency, the promotion of sustainable forms of agriculture, RES implementation, cooperation with other parties, such as the exchange of experience and information, the coordination of national policies, etc.

During the period 1990–2007, the European Commission has registered a reduction of emissions, 7 % in the energy sector, 11 % in the industrial process, 11 % in agriculture (reduced use of fertilizers), and 39 % in the waste sector.

The results of the EU should be interpreted not only in the light of Kyoto and subsequent agreements. The emission reductions are also attributed to two factors: the global economic and financial crisis which has reduced industrial production, and the new member states that joined the EU which have reduced the average level of emissions because of their lower economic productivity compared to the old member states.

1.4 Trends in Global CO_2 Emissions

All the agreements made so far on energy and the environment at global and European level have led today to a certain level of CO_2.

During the last decades, the massive use of fossil fuels and related polluting emissions has made the environmental situation very difficult. The relationship between temperature and CO_2 concentration in the atmosphere has long been an object of study, but today it is confirmed by observation of the long-term cycles. So, the relationship between temperature rise and CO_2 emissions has always been a direct relationship. The difference from the past is that currently the amount of CO_2 in the atmosphere is growing exponentially, as it is shown in Fig. 1.1.

In 2013, the CO_2 concentration has reached almost 400 ppm. During the last glaciations, the concentration of CO_2 in the atmosphere was equal to 180 ppm; in 1800 (preindustrial), it was 280 ppm, and now it has reached 400 ppm. A concentration of 500 ppm is considered by scientists to be a point of no return, where the Earth will stabilize at a new hotter equilibrium. Since 1975, the CO_2 concentration in the atmosphere has increased from about 330 to 400 ppm due to the growing use of fossil fuels (Fig. 1.3). The CO_2 remains in the atmosphere for a very long time and the emissions from any place combine in the whole atmosphere in the period of about 1 year.

Regarding the recent global trend, average monthly global concentration of CO_2 in the atmosphere in 2013 stood at 366.7 ppm, against 393.8 ppm in 2012. Since 2010, the average monthly increases: the atmospheric CO_2 concentration was 389.7 ppm, that is, 7 ppm less than at present.

As the source of more than two thirds of global GHG emissions, the energy sector is crucial to tackling climate change. The growth in primary energy consumption provided by the International Energy Agency (IEA 2014a) in the *World Energy Outlook 2014* (WEO-2014) is shown in Table 1.2 for the period 1990–2040.

Data refer to "New Policy Scenario" (the central scenario in WEO-2014) that explores the evolution of energy markets on the basis of the continuation of existing policies and those implemented until mid-2014.

In particular, the WEO-2014 forecasts the following issues between 2012 and 2040:

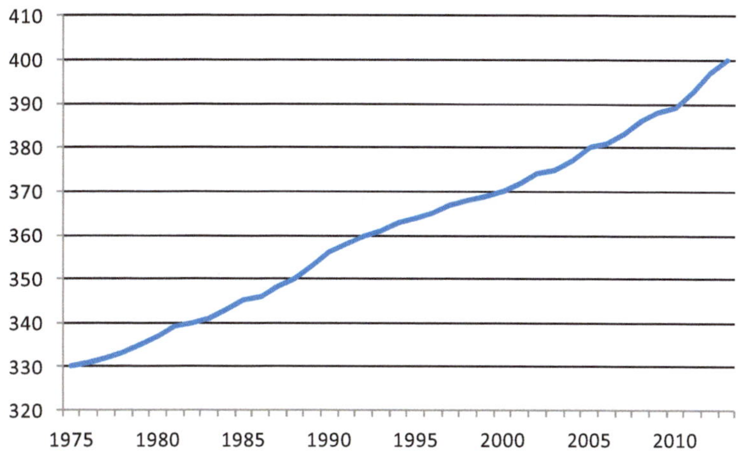

Fig. 1.3 Average CO_2 atmospheric concentration (ppm; 1975—December 2013). (Source: U.S. Department of Commerce 2014)

- The primary energy demand will grow by about 37 %, while the demand for electricity will grow by about 55 %.
- 60 % of the growth in demand for primary energy will be absorbed by China, India, and the Middle East.
- There will be a modest growth in primary energy consumption in the Organisation for Economic Co-operation and Development (OECD) countries where there will be a shift from oil and coal towards gas and RES.
- The highest growth in world primary energy demand will still be due to fossil fuels that will dominate the energy mix in 2040. Their percentage contribution will decrease from 82 to 80 %. Oil is the single largest energy source throughout the period 2012–2040, with a share of 26 % in 2040.
- The two thirds of the expected growth in oil production will come from new technologies: half will come from unconventional techniques in North America (which will allow the USA to become the world's largest producer of oil) and one-fifth extractions from extractions at great depths from floating platforms in Brazil. The report predicts that by 2025, the US production of oil will begin to decline and that oil world market once again will be dominated by the Middle East.
- Gas is among the fossil fuels that will have the greatest growth and its consumption will approach those of coal and oil. Its global demand should increase by 57 % between 2012 and 2040, i.e., 1.6 % per year. Today, North America is increasing gas supply with a consequent reduction of its cost on the world market is divided into separate areas (in the EU, price of gas is about three times that of the USA). The WEO-2014 predicts that both the liquefied natural gas ships and terminals will reduce the differences until there will be a single gas market as it is today that of oil. The wide availability of gas in North America will reduce, in that area, investments in nuclear power plants and to close some coal-fired plants.

Table 1.2 World primary energy demand—expected growth to 2040. (Source: IEA 2014a)

Energy demand (Mt)								CAAGR[a] (%)
	1990	2012	2020	2025	2030	2035	2040	2012–2040
Coal	2231	3879	4211	4293	4342	4392	4448	0.5
Oil	3232	4194	4487	4612	4689	4730	4761	0.5
Gas	1668	2844	3182	3487	3797	4112	4418	1.6
Nuclear	526	642	845	937	1047	1137	1210	2.3
Hydro	184	316	392	430	469	503	535	1.9
Bioenergy	905	1344	1554	1675	1796	1911	2002	1.4
Other renewables	36	142	308	435	581	744	918	6.9
Total primary energy demand	8782	13,361	14,978	15,871	16,720	17,529	18,293	1.1

[a] Compounded average annual growth rate

- The use of coal is expected to grow globally for the contribution of non-OECD countries, with particular reference to China, which in 2012 is the biggest consumer, producer, and exporter at global level. Coal demand should be reduced by 25 % in the OECD countries by 2040 as a result of environmental laws, even if the introduction in the market of the American coal of good quality and low price has increased its use in Europe, especially in Germany, to replace the deficit production due to the closing of nuclear power plants.
- The unconventional techniques for the extraction of oil and gas, according to the WEO-2014's forecasts, will not expand outside North America, although promising sites exist in many areas of the world. The report reckons that outside North America, it will be difficult to overcome the opposition of people. The opposition of the population is based on environmental impacts and legislative differences: in North America, the ownership of the subsoil belongs to whoever owns the land which has an interest to exploit it, while in the rest of the world, the subsoil belongs to the state that might expropriates the land, raising the opposition of the population that is considered damaged by the exploitation of mineral deposits.
- The growth of RES will be approximately three times higher than that occurred in the period 1990–2012, and in 2040, they could provide about 23 % of the electricity consumed in the world against the current 3.3 % (mainly comes from hydroelectric). Regarding the growth forecasts of RES, hydroelectric and wind power are the sources that will grow most even if hydropower will not grow in Europe, Japan, and the USA because the best sites are already exploited. The wind sector will have a strong increase since its cost is now becoming more and more competitive with coal and nuclear power. It will also increase solar energy thanks to many subsidies that governments deliver. For the other RES, the report predicts a considerable development of biomass and biofuels.
- Nuclear power is also growing for power plants planned in China, India, Korea, and Russia.
- The world projected energy consumption will require investments in energy supply for around US$ 40 trillion from 2014 to 2035 (IEA 2014b).

The six largest emitting countries in 2012 are China (29 % in global CO_2), the USA (15 %), the EU (11 %), India (5 %), the Russian Federation (5 %), and Japan (4 %). Total CO_2 emissions from all OECD countries account for one third of global emissions that is the same share as that of China and India. A comparison between national GDP and CO_2 emissions shows there is a strong relation between them, as highlighted in the literature through the Kaya identity which expresses CO_2 emissions from fossil fuel combustion in terms of output levels, energy intensity, and carbon intensity of the output (Raupach et al. 2007). Indeed, the CO_2 emission trend mainly reflects energy-related human activities which are determined by economic growth, particularly in developing countries. Only recently, and mainly in developing countries, a decoupling of the increase in CO_2 emissions from economic growth is taking place thanks to the increased energy efficiency and the implementation of RES (see Table 1.3).

Table 1.3 World balance: expected energy-related CO_2 emissions (1999–2040). (Source: IEA 2014a)

CO_2 emissions (Mt)								Shares (%)		CAAGR[a] (%)
	1990	2012	2020	2025	2030	2035	2040	2012	2040	2012–2040
Coal	8316	13,926	15,081	15,270	15,325	15,396	15,523	44	41	0.4
Oil	8815	11,229	11,811	12,101	12,294	12,417	12,489	36	33	0.4
Gas	3807	6460	7311	7999	8672	9351	10,024	20	26	1.6
Total CO_2	*20,938*	*31,615*	*34,203*	*35,370*	*36,291*	*37,163*	*38,037*	*100*	*100*	*0.7*

[a] Compounded average annual growth rate

In recent years, most developed countries have experienced a recession from which they are recovering slowly. The USA, Canada, and Japan have shown a GDP growth of 2 %, and in many EU countries, economic growth has been even smaller.

Analyzing in particular the EU, it has remained in recession in 2012 with a GDP which declined by 0.3 % compared to 2011 levels (in public–private partnership (PPP) units, IMF 2013). They have been decreased, CO_2 emissions, by 2.3 % in 2012 compared to 2011 levels. The main cause of the decline is primarily emissions reduction in the primary energy consumption of oil and gas by, respectively, 4 and 2 %. These reductions are attributable to the reduction of 1 % in electricity production and emissions reduction in residential and service sectors. Emissions are decreased as well in the transport sector due to a reduction of air and road freights. There has been emissions reduction for 2 % in the energy manufacturing industries which both participate in the EU ETS.

While the total CO_2 emissions from power generation in the EU27 decreased by 2.3 % in 2012, the situation is very different among member states concerning the use of coal, since it has became attractive again for electricity production in EU's energy portfolio.

Between 2011 and 2012, there has been an increase of coal consumption in the UK (+24 %), Spain (+24 %), Germany (4 %), and France (+20 %), and on the other hand, a decrease in Poland (−4 %) and the Czech Republic (−8 %). Most of the coal consumption in these countries is allocated to power generation. In recent years, Germany has shifted production from nuclear power to RES, but it continues to subsidize coal mining until 2018. New coal-fired plants also become operational in Italy (1980-MW plant operating since 2008) and Bulgaria (675-MW plant operating since 2008).

According to the WEO-2014 New Policy Scenario, CO_2 emissions per capita decline slightly on a global basis, and in the EU, emissions should decline by 1.4 % in 2040 (Table 1.4).

1.5 Conclusions

From the analysis carried out so far on energy and environmental policies of the EU in the fight against climate change, we offer important conclusions. To achieve the goal of reducing EU GHG emissions by 80 % by 2050, the road map indicates that a smooth transition in terms of cost requires a reduction in domestic emissions by 40 % in 2030 and 80 % in 2050 compared to 1990 levels. The analysis reveals that existing policies will allow the EU to achieve a 20 % reduction in domestic GHG emissions and the share of EU energy consumption produced from RES to 20 % by 2020.

If the energy efficiency plan would be fully and effectively implemented, thus achieving the goal of a 20 % increase in energy efficiency, the EU would be able to exceed its target of 20 % reduction by 2020.

Table 1.4 Expected EU energy-related CO_2 emissions between 2011 and 2035. (Source: IEA 2014a)

CO_2 emissions (Mt)								Shares (%)		CAAGR[a] (%)
	1990	2012	2020	2025	2030	2035	2040	2012	2040	2012–2040
Coal	1732	1137	960	829	653	493	415	33	18	−3.5
Oil	1656	1408	1263	1162	1055	963	873	41	37	−1.7
Gas	663	898	924	970	993	1026	1048	26	45	0.6
Total CO_2	*4051*	*3442*	*3147*	*2961*	*2701*	*2481*	*2336*	*100*	*100*	*−1.4*

[a] Compounded average annual growth rate

The EU, accounting for just over 10% of global emissions, will not be able to fight alone against climate change. International action is the only way to solve this problem, and the EU must continue to engage its partners. Committed to years in the formulation and implementation of internal policies relating ambitious climate, the EU can drag itself with a number of other countries. At the Copenhagen Climate Change Conference in 2009, world leaders were asked to negotiate a new treaty to bind the entire international community to take measures to limit global warming to no more than 2 °C in 2050 above preindustrial levels. The agreement replaces the Kyoto Protocol which has expired in 2012. To deliver on these commitments, some countries have to implement stronger action than those currently envisaged.

The EU should use this opportunity to maintain environmental leadership and to strengthen cooperation with its international partners, also in order to progressively develop a global carbon market that allows to support the efforts made both by the developed and developing countries to implement strategies of low emission intensity. The current energy road map 2050 strategy represents the continuation of the cycle of the Lisbon Strategy (which has ended in 2010), that is, the EU reform strategy of the last decade that has helped the EU to face the recent crisis. In March 2000 in Lisbon, the EU agreed on a strategic goal for the EU for the years 2000–2010 in order to strengthen employment, economic reform, and social cohesion in the context of knowledge-based economy. With the Lisbon Strategy, the EU set the goal of becoming the most competitive knowledge-based economy in the world, by concentrating efforts towards the two main objectives of sustainable economic growth and employment. The energy road map 2050 strategy is based on the accomplishments achieved so far in the form of partnership for growth and jobs, and it differs from the Lisbon Strategy since it faces new challenges. The new program is the program of all EU member states, large and small, new and old, more or less developed. In fact, the enlarged EU is characterized by different levels of development and therefore different needs; for this reason, the EU 2020 strategy can be modulated as a function of the starting points and different national circumstances in order to promote growth. EU priorities that are a knowledge-based growth as a factor of wealth, involvement of citizens in a participatory society, and competitive and greener economy may be the stimulus factors in the world, and the EU, although small compared to the rest of the world, can act as a guide.

1.6 Appendix 1 The Kyoto Protocol

The EU, accounting for just over 10% of global emissions in 2012, is not able to fight alone against climate change. International action is the only way to solve this problem.

Most of the environmental problems are not national in scope; when more than one country is involved, the typical solution is a treaty that generally comes from international organizations. Among these, it is worth to remember the Intergovernmental Panel on Climate Change (IPCC) which is part of the United Nations (UN).

This organization has been mainly engaged in two activities such as the publication of a five-year assessment report on global emissions of GHG with an evaluation of their impact on the climate and a constantly updated publication of IPCC Guidelines for the assessment of GHG emissions. Another important element of international cooperation in climate, environment, and development is represented by the UNFCCC. The convention has been trying to decide on a level of stabilization of GHG emissions in a way that would prevent dangerous interference with human life, ensuring sustainable economic development. The convention, which entered into force on 21 March 1994, has been ratified by 190 of the 195 states recognized by the UN (signatory countries are defined parties). The main obligations on the parties with respect to the UNFCCC have focused on the estimation and reporting of GHG emissions, their reduction within certain levels and within certain times, and the promotion of access to relevant information on the relating subject. The convention divides countries into three main groups according to differing commitments, and they are listed as Annex I parties, Annex II parties, and non-Annex I parties (Conference of Parties 1997).

Annex I parties include the industrialized countries that were members of the OECD in 1994, plus the Russian Federation, Ukraine and Belarus, and the Central and Eastern Europe countries defined as economies in transition. Annex II parties are Annex I countries excluding countries with economies in transition. These countries are required to provide financial resources to developing countries to help them to reduce emissions and respond to climate change.

The non-Annex I parties are all signatory countries of the convention that are not part of Annex I. The distinction between these groups is not only bureaucratic, but respond to roles, mechanisms, and different instruments. The Conference of the Parties (COP) is the supreme governing body of the UNFCCC. The formal objective of the COP focused on the evaluation of the implementation of the convention, resolutions about the development of new standards of the COP, and the negotiation of new commitments. All binding policies and the relating methods of implementation, including the Kyoto Protocol, have been formally approved by the COP.

The UNFCCC and the COP have been supported by an international secretariat that provides practical and organizational supports, managing all the information related to the convention, including information direct to the public, through a database accessible via web. The work of the COP is supported by two subsidiary bodies that are the Subsidiary Body for Scientific and Technological Advice, a subsidiary body of scientific and technical advice, and the Subsidiary Body for Implementation, a subsidiary body of implementation, assisting the COP in the assessment and revision of an efficient implementation of the convention.

Finally, the Global Environment Facility (GEF) was established in 1991 by the World Bank and the environmental and development programs of the UN in order to finance projects in developing countries. Currently, the GEF includes 183 countries that, through the government institutions, the civil society, and the private sector, deal with environmental issues by supporting national initiatives for sustainable development. In particular, the GEF supports projects in biodiversity, land degradation, climate change, etc.

When the UNFCCC was launched, governments knew very well that its provisions were mostly declarative. The main limitation of the convention was the manifestation of commitments too general and protracted over time in an excessive way, without a short-term vision that could give a range of policy and activities to pursue.

In order to overcome this major limitation, the COP proposed regular meetings once a year. The first meeting, called COP1, was held in Berlin in April 1995. On this occasion, the so-called Berlin Mandate was signed, for which they have created the first concrete actions through the establishment of activities implemented jointly, aimed at testing the feasibility of international projects to reduce emissions.

Among the meetings more profitable and famous, there is definitely COP3 which was held in Kyoto, Japan, in December 1997, where the Kyoto Protocol was defined.

At the beginning of the COP3, it was clear from the start that positions of different countries were very different in relation to the percentage of emission reduction which would have been undertaken. The EU had proposed a generalized reduction of 10% with the possibility of reaching 15% if other countries were willing to do the same effort. The USA was in favor of a stabilization of emissions at 1990 levels by the year 2000, that is, without any reduction. Japan was willing to reduce emissions by 5%. The USA, however, were willing to give up the involvement of developing countries in the reduction targets, provided that developing countries at least would have taken voluntary commitments.

For the USA, however, it was unacceptable both the European proposals to reduce emissions by 10–15% and the redistribution of reduction obligations between the different EU member states. The differentiated quotas were finally accepted on condition that this principle was extended to all Annex I countries. The European delegation was also discordant on a massive use of flexible mechanisms over a maximum of 50% of the reduction commitments of each country. These clashes resulted in a differentiation of commitments in which the so-called group Juscanz (Japan, USA, Canada, Australia, New Zealand) were contrary to defined commitments, for a greater involvement of developing countries and a massive use of flexible mechanisms. The EU was instead in favor of further commitments for industrialized countries. After long and tiring negotiations, the Kyoto Protocol was adopted on the evening of the last day of meetings with a compromise quantified in 5.2% overall emissions and variable shares for the countries of Annex I.

The USA insisted that they should have been relatively less burdened with emission quotas with respect to European quotas, as Congress was hostile to any formal commitment.

Consisting of 28 articles, the Kyoto Protocol has placed a greater emphasis on the objectives assigned to Annex I parties to reduce emissions clearly qualified and quantified for each individual country. The reference period for the implementation of the planned reductions would have to be 5 years, the so-called commitment period from 2008 to 2012. The end result of the protocol, however, was a treaty not well structured and unclear, in which most of the text focused on secondary issues while some key provisions were set out in subordinate sentences.

The moment of entry into force of the Kyoto Protocol was defined by the treaty, that is, when at least 55 countries, including the parties included in Annex I in which the total CO_2 emissions account for 55 % of total emissions in 1990, had deposited their instruments of ratification, approval, acceptance, and accession.

In the fall of 2004, Russia voted in favor of the Kyoto Protocol, thus allowing the entry into force of the protocol 3 months later, on 16 February 2005. Even Australia, with the COP13 in Bali in December 2007, ratified the Kyoto Protocol. Currently, 184 nations have signed the Kyoto Protocol. As important is that the USA, despite being signatories to the treaty, has not yet ratified the same. The Kyoto Protocol establishes clear and specific commitments regarding the limitation of emissions of GHG emissions for the Annex I parties of the UNFCCC.

The gases considered primarily responsible for the greenhouse effect and referenced by the Kyoto Protocol are four:

- CO_2, it is the best known among polluting gases, and it is responsible for 70 % of the greenhouse effect. During the industrial revolution, its concentration has increased by about 30 %, particularly in the Northern Hemisphere. It is produced from natural sources as in the case of the decomposition of plants and volcanic eruptions, and it is partially removed by water and the plants themselves, by means of photosynthesis; it remains in the atmosphere for about a 100 years. In addition, it is also produced by activities such as the use of means of transportation, home heating, electricity production, during which high amounts of fossil fuels are burned. To such emissions, it adds the effect of deforestation, which causes the release of CO_2 and reduces the surface absorption forest.
- Methane, it is less present than CO_2 and is responsible for 20 % of the greenhouse effect. It is mainly produced during the decomposition of organic materials, from dumps and from the biological activity of the animals. It remains in the atmosphere for about 11–12 years.
- Hydrofluorocarbons, their concentration is very low, but their ability to retain heat is from 3000 to 13,000 times higher than that of CO_2; they remain in the atmosphere for up to 400 years.
- Nitrous oxide, it is present in very small part, but it is almost 300 times more potent than CO_2 in retaining heat. It is derived primarily from microbiological processes of nitrification and denitrification, and it is partially absorbed by the oceans.

The objectives set for each country can be found in Kyoto Protocol's Annex B. After the signing of the Kyoto Protocol in 1997, it was thought that Annex I parties would immediately have defined and implemented policies aimed at the progressive reduction of emissions to the 2010 targets. Meanwhile, the period has passed without any significant progress in reducing emissions. Even in countries such as Germany and the UK, where there have been reductions in CO_2 emissions, it was mainly due to economic reasons instead of a careful government's planning.

The protocol requires all Annex I parties to implement policies to improve energy efficiency, as well as reducing emissions. It is also necessary to adequately respond to climate change with appropriate technologies, such as energy efficiency and RES production.

Below there are briefly set out the main provisions and the most important instruments of the Kyoto Protocol.

The Kyoto Protocol has established that Annex I parties should achieve a national GHG inventories system for estimating emissions caused by economic activities by January 2007. The inventories are prepared on the basis of so-called activity data (such as industrial production, cultivated areas, etc.).

The Kyoto Protocol sets that only the methods defined by the IPCC have official status for the preparation of inventories. The preparation of the inventories is mandatory and must include data on emissions in 1990 and each year thereafter, up to 2 years prior to their communication.

In particular, the inventories of Annex I countries must cover the emissions and removals of GHG listed in the Kyoto Protocol, and they must also contain information on GHG such as carbon monoxide, nitrogen oxides, sulfur oxides, and non-methane volatile organic compounds.

Annual inventories of Annex I countries include two parts:

- The common reporting format (CRF), standardized formats for the transmission of inventories as a series of tables and calculation containing only numeric data
- The national inventory report, explanatory reports on methodologies used for the compilation of inventories on sources of data and information on the institutional structures that have followed in the preparation of inventories and procedures of quality control that have been applied.

The national GHG inventories system also facilitates international processes of verification, in which estimates of GHG are subject in order to review its compliance with the properties of transparency, consistency, comparability, completeness, and accuracy in the implementation, as explicitly requested by the UNFCCC and IPCC Guidelines.

The UNFCCC and the Kyoto Protocol require parties to periodically prepare reports, called national communications. These reports must contain information and assessments on emissions and must describe national policies in relation to climate change.

The requirements for the preparation of reports range from Annex I countries to non-Annex I.

To the former, it requires accuracy in the description of the measures envisaged to achieve the targets. Annex I countries must also prepare reports with all the aid provided to non-Annex I.

In contrast, the non-Annex I countries must communicate in their national reports only the types of sources that cause emissions of GHG, along with general comments on the state of economic policy.

Moreover, the Kyoto Protocol establishes that both inventories and national communications of Annex I parties should be subject to checks by review groups established for that purpose. These groups prepare the so-called compilation and synthesis reports on national communications.

Another important institution of the UNFCCC is the Compliance Committee, whose purpose is to evaluate and approve the statements submitted by Annex I parties regarding their emission reductions.

It must be emphasized that in the absence of real and effective sanctions for non-compliance, the countries which do not fulfill the obligations of the Kyoto Protocol undergo only the disadvantage of less favorable measures provided in the protocol itself for some years to come.

So the effective implementation of the protocol is entrusted to the will of the parties.

The verification of fulfillment of the obligations reduction of emissions according to the Kyoto Protocol, in particular with reference to the flexible mechanisms, is carried out with a system of emission allowance or emission units. At the beginning of the commitment period, each party of Annex B receives a certain assigned amount of emission units from the UNFCCC. These units are issued and managed as records on computer systems, called registers. Each unit of output corresponds to 1 t of CO_2 equivalent of GHG. The assigned amount represents total GHG emissions allowed to the parts in relation to their targets. Each country decides for itself whether and how to distribute the amount allocated among its various emission sources.

The Kyoto Protocol introduces the possibility for Annex I countries to fulfill their obligations to reduce GHG emissions through the flexible mechanisms. These instruments represent the clearing of the traditional system of command and control in favor of a market-based approach designed to combine environmental protection operations and profit needs of the industrial world. It seeks, therefore, to contribute to the reduction of the GHG effect through an environmental policy that is not too burdensome for the global system: no more strict obligations and fixed costs, but flexibility and profit margins.

In this light, flexible mechanisms constitute an intervention to reduce emissions additional to national actions taken to fulfill the commitments of the Kyoto Protocol, and they are:

- JI
- CDM
- ETS

The JI mechanism is provided by Article 6 of the Kyoto Protocol. It is a mechanism based on projects to reduce GHG emissions according to which each Annex I country can contribute to the fulfillment of its obligations to reduce emissions through the use of ERUs, resulting from projects undertaken in another Annex I country, in order to limit emissions or increase removals of GHG by so-called carbon sinks. The ERUs may be issued only for the period 2008–2012. The country investor may also involve other legal units, such as industry or institutional investors, in transactions of transfer and acquisition of emission reductions, pledging anyway to control their actions in the international context.

The CDM is provided by Article 12 of the Kyoto Protocol which defines the general aspects, without providing a complete discipline. This mechanism is configured as a mechanism very similar in its basic structure to the JI, since it too is based on projects that reduce GHG, with the only difference that the plans for the reduction of emissions are made in a non-Annex I countries, that is, without explicit commitments of communication and thus of emission reductions.

The importance of the CDM is the promotion of sustainable development given its dual purpose, that is, from one hand to allow Annex I countries to promote and implement projects of GHG emission reductions in the non-Annex I in order to use the resulting emission reductions to help meeting their obligations under the Kyoto Protocol, and from the other hand to allow countries to non-Annex I host projects that reduce emissions of GHG and benefit from environmental and social issues related to the transfer of clean technologies and to improve their energy efficiency.

The ETS allows to trade quotas for GHG emissions among countries of Annex B that are OECD member countries already included in Annex I. Through this mechanism, countries that emit less than their assigned amounts have the possibility to sell surplus emission units to other parties whose emissions, on the contrary, exceed the targets. The problem within the ETS is that almost all Annex B countries exceed their emission limits, and so basically, there is no excess allowances.

References

Bigerna S, Polinori P (2014) Italian households' willingness to pay for green electricity. Renew Sustain Energy Rev 34:110–121

Borghesi S (2010) The european emission trading scheme and renewable energy policies: credible targets for incredible results? Nota di Lavoro 141.2010, Fondazione Eni Enrico Mattei

Carraro C, Eychmans J, Finus M (2006) Optimal transfers and participation decisions in international environmental agreements. Rev Int Organ 1:379–396

Conference of the Parties (1997) Kyoto Protocol to the United Nations framework convention on climate change. Third Session Kyoto, 1–10 December

Cremer H, Gahvari F (2002) Imperfect observability of emissions and second best emission and output taxes. J Public Econ 85(3):385–407

European Commission (1996) Council Directive 96/61/EC of 24 September 1996 concerning Integrated Pollution Prevention and Control (IPPC). Official J Eur Union L 257:26–40

European Commission (1999) Council decision 1999/296/EC of 26 April 1999 amending decision 93/389/EEC for a monitoring mechanism of community CO_2 and other greenhouse gas emissions. Off J Eur Union L 117:35–38

European Commission (2001) Directive 2001/77/EC of the European Parliament and of the Council of 27 September 2001 on the promotion of electricity produced from renewable energy sources in the internal electricity market. Off J Eur Union L 283:33–40

European Commission (2002a) Council Decision No 358/2002/EC of 25 April 2002 concerning the approval, on behalf of the European Community, of the Kyoto Protocol to the United Nations Framework Convention on Climate Change and the joint fulfilment of commitments there under. Off J Eur Union L 130:1–3

European Commission (2002b) Council decision no 1600/2002/EC of the European Parliament and of the Council of 22 July 2002 laying down the Sixth Community Environment Action Programme. Off J Eur Union L 242:1–15

European Commission (2003) Directive 2003/87/EC of the European Parliament and of the Council of 13 October 2003 establishing a scheme for greenhouse gas emission allowance trading within the Community and amending Council Directive 96/61/EC. Off J Eur Union L 275:32–46

European Commission (2004) Directive 2004/101/EC the European Parliament and of the Council of 27 October 2004 amending Directive 2003/87/EC establishing a scheme for greenhouse gas emission allowance trading within the Community, in respect of the Kyoto Protocol's project mechanisms Text with EEA relevance. Off J Eur Union L 338:18–23

European Commission (2009) Directive 2009/28/EC of the European Parliament and of the Council of 23 April 2009 on the promotion of the use of energy from renewable sources and amending and subsequently repealing Directives 2001/77/EC and 2003/30/EC. Off J Eur Union L 140:16–62

European Commission (2011a) Communication (2011) 21 from the Commission to the European Parliament, the Council, the European Economic and Social Committee and the Committee of the Regions. A resource-efficient Europe—Flagship initiative under the Europe 2020 Strategy

European Commission (2011b) Communication (2011) 112 from the Commission to the European Parliament, the Council, the European Economic and Social Committee and the Committee of the Regions. A Roadmap for moving to a competitive low carbon economy in 2050

European Commission (2014) Communication (2014) 0015 from the Commission to the European Parliament, the Council, the European Economic and Social Committee and the Committee of the Regions. A policy framework for climate and energy in the period from 2020 to 2030

Eurostat (2014) http://epp.eurostat.ec.europa.eu/portal/page/portal/statistics/themes. Accessed 9 Jun 2014

Grubb M, Neuhoff K (2006) Allocation and competitiveness in the EU emissions trading scheme: policy overview. Climate Policy 6(1):7–30

International Energy Agency (IEA) (2010) World Energy Outlook 2010, OECD/IEA, Paris

International Energy Agency (IEA) (2014a) World Energy Outlook 2014, OECD/IEA, Paris

International Energy Agency (IEA) (2014b) World Energy Investment Outlook 2014, OECD/IEA, Paris

International Monetary Fund (IMF) (2013) http://www.imf.org/external. Accessed 12 Jun 2014

Kolstad CD (2000) Environmental economics. Oxford University Press, Oxford

Raupach MR, Marland G, Ciais P, Le Quéré C, Canadell JG, Klepper G, Field CB (2007) Global and regional drivers of accelerating CO_2 emissions. Proc Natl Acad Sci USA 104(24):10288–10293

Stavins RN (1995) Transaction costs and tradeable permits. J Environ Econ Manage 29(2):133–148

U.S. Department of Commerce–Earth System Research Laboratory (2014) http://www.esrl.noaa,gov/gmd/ccgg/trends.html

Chapter 2
RES Scenario in the European Union

The evolution of energy consumption in Europe is going through a profound change in recent years: the incidence of traditional fuels is diminishing gradually in the face of an increase in renewable energy sources (RES). In 2013, the consumption of electricity from RES amounts to 14 % of the total with respect to 6 % in 1997. This chapter provides a quantitative analysis of RES in the EU and the prospect of development of RES under different scenarios. The analysis begins with a description of the technical characteristics of RES and the quantity produced by each source at world level. In Sect. 2.2, the distribution of the three major RES that are wind, solar and biomass throughout the EU context is analyzed. Section 2.3 provides an analysis of the development of RES in the coming years by different scenarios, both according to the 2014 edition of future scenarios of the International Energy Agency (IEA 2014a) in the World Energy Outlook-2014 (WEO-2014) and the current EU climate and energy package. Conclusions are drawn in Sect.clus.

2.1 Power Generation from RES

In recent years, especially after repeated threats that affect the environment, RES have acquired significant positions in public opinion and among energy experts. This is because they have as main characteristic the renewability, i.e., the ability to provide energy without exhausting over time. In fact, RES are one of the engines of sustainable development because they do not affect the storage of energy resources and allow to produce cleaner energy. RES have a huge amount of advantages that affect political, economic, and ethical issues. Their use involves a reduction in greenhouse gas (GHG) emissions, greater security supply of primary energy sources, the stabilization of energy costs in the long run since RES have raw material costs in only a few cases (e.g., biomass), high conversion efficiency (thermal and electric) due to technological innovations. Next to the great advantage of producing cleaner energy, RES also show some drawbacks. The production of RES is intermittent in time due to variability in daily, seasonal, climatic conditions of the primary source and there are high investment costs. Systems of the various RES require very large

S. Bigerna et al., *The Sustainability of Renewable Energy in Europe*,
DOI 10.1007/978-3-319-12343-1_2

land areas and these can have a little look natural both for the shape of the components and the visual impact.

The energy sources are grouped into two macro groups: sources relating directly or indirectly to solar radiation and sources relating to materials or phenomena that occur on earth independently of solar irradiation. Among the sources derived directly or indirectly from solar radiation there are the kinetic energy associated to the flow of the winds, the kinetic energy associated with ocean currents deep, biomass, photovoltaic (PV), energy connected to swell, and the chemical energy contained in fossil fuels.

Among the sources derived from materials or phenomena that occur on earth independently of solar radiation there is the energy contained in the bonds between intra-atomic particles and the kinetic energy related to the motion of the tides and the motion of the current surface ocean, geothermal energy.

The Directive 2001/77/EC (European Commission 2001, 2011) states that:

- RES, renewable nonfossil energy sources (wind, solar, geothermal, wave, tidal, hydropower, biomass, landfill gas, residual gas from sewage treatment, and biogas);
- biomass, means the biodegradable fraction of products, waste, and residues from agriculture (including vegetal and animal substances), forestry, and related industries, as well as the biodegradable fraction of industrial and municipal waste;
- electricity produced from RES, electricity produced by plants using only RES, as well as the share of electricity produced from RES in hybrid plants also using conventional energy sources, including renewable electricity used for filling storage systems, and excluding electricity produced as a result of storage systems;
- electricity consumption, the national electricity production, including autoproduction, plus imports, minus exports (gross national electricity consumption).

Features common to most RES that have direct implications in the power system are the variability, the location, modularity, uncertainty on the intensity of the source, and low operating costs since no fuel costs occur.

Here follows a brief and nontechnical description of the main sectors of intervention and investment in electricity generation plants fueled by RES, by the type of RES, particularly with regard to hydropower, geothermal, wind, solar PV, and biomass because of their greater industrial importance compared to other RES.

The development of current wind technology began with the oil crises that hit particularly industrial countries from 1973 onwards. Wind was considered as one of the most promising alternatives to fossil fuels for electricity production. At the beginning of the 1980s, current wind technology was already tested in the context of government programs in Denmark, Britain, Germany, Sweden, and the USA, several prototypes of wind turbines with power up to 4 MW. The progressive introduction of incentives in various countries, especially in Europe, helped to keep alive the demand. At the beginning of the 1990, European, American, and Japanese companies came to produce machines with power ratings up to 300 kW. In more recent years, machines have been placed in the market wind turbines with power ratings between 1.5 and 3 MW. Today, the wind farm built on land in Europe typically has a total power of tens of mega watt.

Offshore installations[1], as well as terrestrial ones in countries with wider availability of space, can be up to several 100 MW each.

Installations of wind power generation in use today can be divided into two categories:

- the wind farm power significantly related to the medium-voltage networks and especially the high voltage;
- the mini wind turbines, that is low power plant (not more than 100–200 kW) used for feeding of users not served by the network with the help of accumulation systems or for connection to low-voltage networks.

The key component of both types of system is the machine that converts wind energy into electricity, i.e., the wind turbine. In wind farms, a certain number of wind turbines are installed in patterns appropriate in an area with adequate wind conditions. The power of each unit today is tend to exceeding 1 MW. The mini wind turbines instead often include a single wind turbine whose power is 1 kW (or less) up to 100–200 kW.

The division of the cost of a wind farm in a typical land site is on average 70–80 % of the purchase and installation of the wind turbines and 20–30 % for the remainder, including the foundations of the wind turbines, the streets of access, internal roads, electricity infrastructure, installation work, the line connecting to the network, and the planning and authorization procedures. The incidence of these costs is naturally higher in hilly or mountainous areas.

Generally, the typical costs of plant lie in the range between 1300 and 2000 €/kW, with the highest values for relatively small plants in remote areas in complex terrain.

The offshore farms have a much higher installation cost compared to terrestrial ones, both for the higher prices of wind turbines designed for offshore use and for the higher costs of foundations underwater, installation in the sea, and electrical infrastructure connection to the mainland. For the most recent projects, estimates range from 2800 to almost 4000 €/kW; the ma higher costs of offshore should be offset by higher annual energy production.

In order to calculate the average cost of a wind power plant project, it is necessary to evaluate its annual net real manufacturability and it is appropriate to make assumptions regarding the useful life of the machines, usually 20 years, and the discount rate. For instance, for an average Italian wind farm of 20 MW with specific cost 1740 €/kW for a life of 20 years and a discount rate of 5 %, it is estimated an average cost of between 110 and 160 €/MWh. Of course, in other countries with easier terrain, open spaces, and strong and regular wind regimes (e.g., in the

[1] The offshore wind is the wind made in the sea. Thanks to the favorable conditions of wind at sea, offshore wind (with the same installed power) produces on average 30 % more energy than usual onshore.

north-central EU), the average cost of energy may be much lower, even in the order of 50–60 €/MWh.

The production of wind energy has mainly two controversial aspects. First, the average cost of wind power generation is still above the EU average revenues obtained from the sale of the wind power in the electricity market. It then explains how the wind power production still needs some incentives (it follows a detailed description of incentive mechanisms in Chap. 3).

Second, the development of RES, in particular wind energy, introduces elements of landscape transformation, thus an evolution of the ratio between the population and perception that they have of the landscape itself. The plants powered by RES are deeply rooted in the territory in which they are installed and from which they receive their raw material, making it more visible energy production and transforming the relationship with the natural environment. The presence of environmental impacts related to RES imposes the need for a shared scenario and an integrated approach to the comparison of the overall benefits and the changes induced locally.

The wind power capacity installed worldwide during 2012 is 44,184 MW, corresponding to a cumulative capacity of 281,052 MW worldwide (Table 2.1).

Wind continues to be used mainly by wind power plants of significant power, connected to high-voltage power grids and wind turbines mainly between 1 and 3 MW of power per unit on the mainland and between 2 and 6 MW offshore.

Concerning the PV technology, it allows to directly convert solar energy into electricity through the PV effect, that is the property of some semiconductor materials to generate electricity when struck by light radiation. The silicon element is widespread in nature, is the basic material for the PV cell, and is the elementary device that can produce about 1.5 W in direct current, normally insufficient for common uses. More cells are electrically connected and encapsulated in a structure to form the module-based component which is commercially available.

Several modules are connected in series and in parallel, forming sections of a plant whose power can reach thousands of kilowatt. Downstream of PV modules is placed the inverter that converts the direct current into alternating current generated by the cell, directly usable by users or with the possibility of being fed into the grid. The modules can be oriented towards the sun on fixed structures, or structures able to follow the movement in order to increase solar collection.

Each kilowatt peak[2] installed requires an area of about 8–10 m^2 net for crystalline silicon modules coplanar to the roofs of buildings; it is necessary instead of a broader space for modules arranged in several rows on a flat surface to reduce shading.

The main applications of PV systems are:

- Systems with storage system for isolated users from the network; such systems are independent type and they are indicated in the case of the absence of electrical connection (such as mountains, in agricultural areas not served by the network,

[2] The kilowatt peak is used to measure the maximum power delivered by a photovoltaic generator.

Table 2.1 Global wind capacity (MW). (Source: EurObserv'er 2014 and ENEA 2014)

	Cumulative capacity (2011)	Cumulative capacity (2012)	Installed capacity in 2012
Europe	96,732.8	109,176.1	12,690.0
USA	46,919.0	60,007.0	13,124.0
Canada	5265.0	6200.0	935.0
China	62,364.0	75,564.0	13,200.0
India	16,084.0	18,421.0	2337.0
Japan	2536.0	2614.0	88.0
Other Asian countries	1086.0	1211.0	125.0
Africa and Middle East	1033.0	1135.0	102.0
Latin America	2280.0	3505.0	1225.0
Pacific Region	2865.0	3219.0	358.0
Total	*237,160.8*	*281,052.0*	*44,184.0*

etc.). The stand-alone PV systems are the best solution to address the lack of electricity network. This technology can be very useful in developing countries that have problems of energy supply to hospitals, schools, and kindergartens, and also in remote areas (in the mountains or countryside) not served by the national grid.
- PV plants connected to the grid: they have the feature to work on an interchange with the local power grid. In particular, during daylight hours, the user consumes the electricity produced by his solar system, and when there is no light or it is not sufficient, or if the user requires more energy than the plant can provide, the electricity network will ensure the supply of the electricity needed. On the other hand, in the case, the plant produces more energy than required by the user, such energy can be fed into the network. Solar installations connected to the grid thus represent an additional source, because they provide a contribution for a different amount depending on the size of the plant, the building's overall electricity balance that uses them. There are also power stations connected to the network in the medium or high voltage.
- "Anti-blackout" PV systems: they are hybrid systems which provide electricity during power cuts by taking energy from the battery (as in systems for isolated users), while working as normal systems to connect to the network when there is electricity.

Of all the energy that hits the solar cell in the form of light radiation, only a portion is converted into electric energy available at its terminals. The conversion efficiency of commercial silicon cells is generally between 13 and 20 %, while laboratory cells have reached values of 32.5 %.

A PV system produces electricity for 20–25 years, with few maintenance requirements and a good resistance to atmospheric agents. Disposal at the end of life does not have any particular problems: a PV module is in fact more than 90 % recyclable.

The most significant cost item for the construction of a solar system is certainly made up of the purchase cost of PV modules that affect the total to 40–60 % (depending on power). The PV modules are the most long-lived component of a plant, as designed and manufactured to produce electricity for more than 50 years (some researchers claim that they can work for over 100 years). The design and installation together range from 30 % for smaller plants to about 15 % for the major power plants. The remaining cost items are almost always below 10 %. In particular, the inverter is at most 10 % of the total cost of the system for a power range of 20–100 kWp, which declined gradually to 7 % in the case of plants from 3 kWp. The support structures of the modules instead vary from 5 to 8 % for traditional structures depending on the size of the system.

In 2012 there have been installed 31.4 GW worldwide, an increase of 14 % compared to 2011 and in 2013 the PV has grown more than 13 %. The EU accounts for the predominant share of the global PV market, with 55 % of all new capacity in 2012 and about 70 % of the world's cumulative PV capacity with more than 70 GW of cumulative installed capacity in 2012. China is the second largest producer of PV with cumulative capacity amounts to 8.3 GW in 2012, followed by the USA

Table 2.2 Global photovoltaic cumulative installed capacity (MW). (Source: EurObserv'er and ENEA 2013)

	Cumulative capacity (2011)	Cumulative capacity (2012)	Installed capacity in 2012
Europe	52,884.0	70,043.0	17,159.0
Asia Pacific	7628.0	12,397.0	4769.0
Americas	4959.0	8717.0	3758.0
China	3300.0	8300.0	5000.0
Middle East and Africa	192.0	601.0	409.0
Rest of the world	2098.0	2098.0	0.0
Total	*71,061.0*	*102,156.0*	*31,095.0*

(7.8 GW) and Japan (6.9 GW). Some countries that have a huge potential, as Africa, the Middle East, South East Asia and Latin America, are beginning to appear on PV market.

At the global level, nine of the top ten markets for PV in 2012 have installed at least 1 GW of PV systems (Table 2.2).

Concerning the hydroelectric technology, hydroelectric power plant converts the hydraulic energy of a watercourse, natural or artificial, into electrical energy. In general, the functional scheme includes the work of the barrage, a dam, or a crossbar, which intercepts the course of water creating a reservoir which may be a tank, or basin, where it is kept in an almost constant level of water.

Through the works of water supply, canals, and tunnels, water is then piped into tanks and cargos and through penstocks, it flows in turbines via the inlet valves (safety) and regulatory bodies of the range (distributors) according to energy demand. The water, which puts into action the turbines, then flows out of the discharge channel through which it is returned to the river. The alternator is directly connected to the turbine and is mounted in an arrangement with a vertical axis or horizontal axis. It is essentially a rotating electrical machine capable of transforming mechanical energy into electrical energy by the turbine. The electricity thus obtained must be transformed in order to be transmitted over long distances.

Therefore, before being conveyed in transmission lines, electrical energy passes through the transformer that lowers the intensity of the current produced by the alternator; however, by raising the voltage to thousands of volts. Arrived at the place of use, before being used, the energy passes back into a transformer that this time raises the intensity of current and lowers the voltage so as to make it suitable for domestic use.

The power of a hydraulic system depends on two factors that are the jump (height difference between the height at which the water resource is available flared and the level at which the same shall be refunded after the passage through the turbine) and the rate of flow (the mass of water flowing through the machine expressed per unit of time).

There are also mini hydroelectric plants which are hydroelectric plants with a capacity less than 100 kW. The micro-stations fall into one broader category, called micro-hydraulic by United Nations Industrial Development Organization[3] (UNIDO), a term that indicates hydropower plants with a capacity below 10 MW.

The micro-hydraulics plants are classified as:

- Pico-hydropower plants, $P < 5$ kW;
- Micro-hydropower plants, $P < 100$ kW;
- Mini-hydropower plants, $P < 1000$ kW;
- Small-hydropower plants, $P < 10{,}000$ kW.

Both in the case of large and micro plants, the generation of electricity by hydroelectric presents environmental benefits of not polluting the environment with substances discharged into the air, as it happens in the case of traditional methods of generation by thermoelectric.

The micro- hydropower plants mainly solve the problems associated with large systems: from more parts, large dams are under indictment for the damage they might cause to natural ecosystems as the variation in the quantity and the quality of water resulting from the construction of dams can have potential impact on fish fauna along with any changes in the riparian vegetation.

In many cases, moreover, the micro-hydropower stations have positive effects on adjustment and retention of floodwaters of the streams, especially in mountainous areas where they often cause instability and degradation of the soil.

Approximately 20 % of the electricity produced in the world today comes from hydro. Hydropower in fact contributes to a significant share of the world demand for primary energy: to a greater extent than 6 % worldwide, but with values much higher for some regions (e.g. 27 % in the case of Central and South America) or for individual countries. Global capacity in 2012 amounts to 990 GW, with China accounting for 23 % with 231 installed capacity, Europe for 15 % (148 GW installed capacity), Brazil 8.5 % (82 GW installed capacity), USA 7.9 % (76 GW installed capacity), Canada 7.8 % (75 GW installed capacity) and Russia 4.6 % (50 GW installed capacity). Globally, hydropower has generated at around 3700 TWh of electricity during 2012 (Table 2.3).

The biomass energy is applied to more than one material, virgin or residual agricultural, or industrial processes, which may occur in different physical conditions.

Depending on the type of biomass and the most appropriate technology for implementing the energy enhancement, there are many different plant solutions.

The plants that are most widely are:

- Conventional, with combustion of solid biomass
- Gas turbine fueled by synthesis gas, obtained from biomass gasification
- Combined cycle, with steam turbine and gas turbine

[3] UNIDO is the specialized agency of the United Nations that promotes industrial development for poverty reduction, inclusive globalization, and environmental sustainability.

Table 2.3 Hydropower capacity in 2012. (Source: World Energy Council 2013)

	Installed capacity (MW)	Actual generation (GWh)
China	231,000.0	714,000.0
Europe	148,500.0	
Brazil	82,458.0	428,571.0
USA	77,500.0	268,000.0
Canada	75,104.0	348,110.0
Russian Federation	49,700.0	180,000.0
Rest of the world	325,738.0	828,437.0
Global total	*990,000.0*	*2,767,118.0*

- Thermoelectric hybrids, that use both biomass and conventional sources
- Powered by liquid biomass (vegetable oils, biodiesel), consisting of engines coupled to generators

Biomass is the most sophisticated form of accumulation of solar energy which through photosynthesis converts light energy into chemical energy stored in organic molecules. For this reason, it is an RES as the CO_2 produced during combustion is reabsorbed by plants through photosynthesis.

Currently, biomass meets about 15 % of primary energy use in the world, with 55 million TJ/year (1230 Mtoe/year). The use of this source shows, however, a high degree of heterogeneity between countries. Developing countries, on the whole, on average, derive 38 % of its energy from biomass, with 48 million TJ/year (1074 Mtoe/year), but in many of them the resource meets up to 90 % of energy needs total, by the combustion of wood, straw and animal waste. In industrialized countries instead biomass contributes for 3 % of the primary energy use by 7 million TJ/yr (156 Mtoe/year).

A widespread use of biomass can result in significant impacts at economic, environmental, and employment levels, as they can ensure enhancement of agro-industrial residues, new opportunities for the development of marginal areas and/or reduction in agricultural surpluses with the replacement of traditional crops with energy crops, very low contribution to the increase in the rate of CO2 in the atmosphere. In order to diversify RES, the use of biomass is an important potential energy reservoir, which could help to reduce vulnerability in the supply of energy resources and restrict the import of electricity for many countries. Biofuel production that is any fuel derived from biomass is shown in Table 2.4.

Geothermal energy uses the earth's internal heat generated in part during the formation of the planet and in part by the decay of radioactive isotopes in the mantle (volcanoes, hot springs, fumaroles, and geysers are indications of the heat that is stored in the Earth's crust). A geothermal power plant consists of an area where it was detected in the presence of hot fluids, with depths ranging from 60 to 3000 m from where the vapor presents at high temperatures (150–250 °C) is extracted by drilling and then piped into insulated carbon steel pipes, which carry the steam to the central.

Table 2.4 Biofuel global production, 2012 (billion liters). (Source: REN21 2013)

	Fuel ethanol	Biodiesel	Total	Comparison with volumes produced in 2011
USA	50.4	3.6	54	−2.4
Brazil	21.6	2.7	24.3	0.6
Germany	0.8	2.7	3.5	−0.5
Argentina	0.2	2.8	3	0.1
France	1	1.9	2.9	0.2
China	2.1	0.2	2.3	No change
Canada	1.8	0.1	1.9	0.2
Thailand	0.7	0.9	1.6	0.5
Indonesia	0.1	1.5	1.6	0.2
Spain	0.4	0.5	0.9	−0.3
Belgium	0.4	0.4	0.8	No change
The Netherlands	0.2	0.5	0.7	−0.1
Colombia	0.4	0.3	0.7	No change
Austria	0.2	0.4	0.6	No change
India	0.5	>0.0	0.5	0.1
EU-27	4.2	9.1	13.3	−0.7
World total	*83.1*	*22.5*	*105.6*	*1*

The energy possessed by the vapor, expanded in a turbine coupled to a generator, is first transformed into mechanical energy and then into electrical energy. The exhausted steam is conveyed into the condenser, where it is converted to water at high temperature, which passes into the cooling tower, where it is cooled and injected underground. It is of fundamental importance to ensure the renewability of the geothermal system, replacing the fluid removed from use, both to prevent landslides and in order not to deplete the geothermal fields. This can occur naturally through the rain, while in other cases it is necessary to artificially recharge through injection wells.

The conversion of geothermal energy into electricity is obtainable with different technologies depending on the temperature and pressure of the hydrothermal system available:

- Dry-steam systems that consist in majority by dry steam which is located at high pressures and temperatures accompanied by other gases or soluble substances. The steam can be used directly for the production of electricity conveying it to a turbine.
- Systems to wet steam: it is constituted by hot water at a temperature higher than its boiling point and high pressure; when pressure is reduced in the column of the well the water vaporizes and comes to the surface in the form of a mixture composed of water and steam. The steam can be used for the production of electricity, while the hot water can be used in desalination plants to produce freshwater. The temperature in this type of system is between 180 and 370 °C.
- Hot-water systems: they contain water at a temperature below 100 °C can be used in most cases for direct uses such as home heating, greenhouses, and industrial plants.

The use of geothermal energy can be divided into some areas: as uses at high enthalpy with fluids that reach temperatures higher than 150 °C; it concerns the production of electric energy and some industrial uses and uses at medium- and low-enthalpy fluids that reach temperatures of 150–100 °C in the first case and less than 100 °C in the second; and it concerns the direct uses as domestic, agricultural, and industrial ones.

The overall yield of the production of electricity from geothermal steam is 10–17 %, about three times lower than that of traditional sources, this is due to the lower temperature of the steam (250 °C), both for the different chemical composition of the same, which determines a loss of energy. Geothermal power stations consume from 6 to 15 kg of steam and a good production well, with a flow rate of 70,000 kg/h, can feed a power of 10 MW. On land, there are large areas in which the subsoil are fluids at temperatures between 40 and 100 °C, easily accessible, which could be directly used for heating and cooling, thus allowing a considerable saving of hydrocarbons.

2.2 RES in the EU

In the EU, in 2012, the energy coming from RES accounted for 14.1 % of gross final consumption, covering about 8 % of EU electricity demand. There is an increase from 1 year to the other of the weight of the share of energy consumed from RES compared to the total energy consumed on average (Table 2.5).

Overall, RES status in the EU is good since some RES technologies have shown rapid progress. In fact, what you thought was in the 1990s an ambitious target for 2010 has already been reached for some technologies. In 1997, the European Commission had made some projections for RES in its White Paper for a Community Strategy and Action Plan. Wind and PV have even exceeded the target: wind energy, with a cumulative installed capacity at the end of 2010, amounting to 84 GW has exceeded the target of the White Paper of 40 GW; PV, at the end of 2010, has shown an installed capacity of approximately 27 GW against the target of 3 GW of installed capacity.

Albeit slowly, the use of RES increased in the EU, reaching a share of 14.1% in the Eu final energy consumption. According to Eurostat (2013), Sweden, Bulgaria, and Estonia have already achieved the 2020 targets set by the RES Directive (2009/28/EC). The EU countries that have increased the consumption of RES are Sweden (increased from 38.7 % in 2004 to 51 % in 2012), Denmark (from 14.5 to 26 %), Austria (from 22.7 32.1 %), Greece (from 7.2 to 15.1 %) and Italy (from 5.7 to 13.5 %). The countries most green in absolute terms are Sweden (EU record with 51 %, exceeding eight years ahead of the target of 49 %), Latvia (35.8 %), Finland (34.3 %) and Austria (32.1 %). Estonia was the first to reach the 2020 target early in 2011 (extendable to 25.2 % in 2012), followed in 2012 also from Bulgaria (16.3 %). The less virtuous countries regarding the use of RES are Malta (1.4 %), Luxembourg (3.1 %), Great Britain (4.2 %) and the Netherlands (4.5 %).

It is interesting to look at the situation of the individual RES in the EU.

With reference to wind power, 2012 has been a positive year for the EU wind energy market: the new installed capacity during the year was 11840 MW enabling the EU to greatly exceed the threshold of 100 GW of installed capacity (MW 105635). Subtracting the disused installations, wind power of the EU has increased by 11,593 MW in 2012, compared with an increase of 9299 MW in 2011.

The wind power capacity in the EU is currently 209.7 kW per 1000 inhabitants. By taking into account this indicator to assess the weight of wind power in a single state, the main EU countries are Denmark (754.8 kW/1000 inhabitants), Spain (4888 kW/1000 inhabitants), and Portugal (429.2 kW/1000 inhabitants).

Several factors have contributed to the strong growth of the EU market in 2012 and mainly they are related to offshore wind farms at high power in the North Sea as well as onshore parks in Scotland. Another factor is the emerging markets of eastern EU countries (Poland, Romania, and Austria in particular) that have been very dynamic in 2012, driven by the sharp increase in the price of gas.

As regards offshore wind, the power connected to the network in the EU amounted to 4705.8 MW in 2012, which corresponds to about 1156.4 MW of additional

Table 2.5 Primary production of RES, 2000 and 2010. (Source: Eurostat 2013)

	Primary production (1000 toe)		Share of total, 2010 (%)				
	2000	2010	Solar energy	Biomass and waste	Geothermal energy	Hydropower energy	Wind energy
EU-27	96,650	166,647	2.2	67.6	3.5	18.9	7.7
Euro area	65,006	118,679	2.9	64.3	4.8	19.0	9.0
Belgium	534	1989	3.0	89.8	0.2	1.4	5.6
Bulgaria	780	1475	0.8	63.6	2.2	29.5	4.0
Czech Republic	1339	2900	2.1	88.6	0.0	8.3	1.0
Denmark	1766	3123	0.5	77.6	0.3	0.1	21.5
Germany	9094	32,746	4.4	78.7	1.6	5.4	9.9
Estonia	512	988	0.0	97.3	0.0	0.2	2.4
Ireland	235	620	1.0	51.8	0.0	8.4	39.0
Greece	1403	1985	9.9	44.7	1.4	32.3	11.7
Spain	6928	14,657	7.0	42.2	0.1	24.8	25.9
France	15,874	20,793	0.5	69.1	0.4	25.6	4.1
Italy	9598	16,328	1.8	37.3	29.2	26.9	4.8
Cyprus	44	77	79.2	15.6	1.3	0.0	3.9
Latvia	1393	2101	0.0	85.4	0.0	14.4	0.2
Lithuania	682	1185	0.0	94.0	0.4	3.9	1.6
Luxembourg	39	92	3.3	81.5	0.0	9.8	5.4
Hungary	830	1922	0.3	91.4	5.2	0.8	2.4
Malta	0.0	0.0	0.0	0.0	0.0	0.0	0.0
The Netherlands	1347	2896	1.0	86.6	0.3	0.3	11.8
Austria	6608	8600	2.0	57.1	0.4	38.4	2.1
Poland	3808	6849	0.0	94.0	0.2	3.7	2.1

Table 2.5 (continued)

	Primary production (1000 toe)		Share of total, 2010 (%)				
	2000	2010	Solar energy	Biomass and waste	Geothermal energy	Hydropower energy	Wind energy
Portugal	3759	5438	1.4	55.1	3.5	25.5	14.5
Romania	4040	5677	0.0	69.6	0.4	29.6	0.5
Slovenia	788	1041	0.6	59.5	2.7	37.3	0.0
Slovakia	496	1398	0.0	67.0	0.6	32.3	0.1
Finland	7748	9030	0.0	87.4	0.0	12.3	0.3
Sweden	14,741	17,408	0.1	65.4	0.0	32.8	1.7
UK	2264	5327	1.7	76.0	0.0	5.8	16.4
Norway	13,481	11,554	0.0	11.9	0.0	87.5	0.7
Switzerland	4437	4968	1.0	31.3	5.2	62.4	0.1
Croatia	879	1232	0.4	39.9	0.6	58.1	1.0
FYR (Former Yugoslavic Republic) of Macedonia	322	422	0.0	47.6	2.8	49.5	0.0
Turkey	10,102	11,627	3.7	38.9	16.9	38.3	2.2

Table 2.6 Cumulative installed offshore wind power capacity in 2012 (MW). (Source: EurObservurO and ENEA 2013)

	2011	2012
UK	1838.0	2679.0
Denmark	817.5	921.9
Belgium	195.0	380.0
Germany	200.3	280.3
The Netherlands	228.0	228.0
Sweden	163.4	163.4
Finland	26.0	26.0
Ireland	25.2	25.2
Portugal	2.0	2.0
Total EU 27	*3549.4*	*4705.8*

power connected to the network compared to 2011. According to the European Wind Energy Association (EWEA), offshore power's increase is mainly due to new plants put into operation in the UK that have brought the cumulative offshore capacity in the UK to 2679 MW at the end of 2012. Denmark is the country's second largest producer of offshore wind, with a total installed capacity in the late 2012 at around 912.9 MW (Table 2.6).

Although the EU wind energy market in 2012 has exceeded expectations, prospects for growth in the short term are less encouraging. Among the causes, the political uncertainty that is characterizing the EU has led a number of countries to review their incentive systems or not to provide sufficient guarantees to investors in the coming years. In fact, during 2013 in the EU, 28 have been installed 11,159 MW of new wind energy capacity, a decrease of 8% compared to 2012. This is what emerges from the report wind power in 2013 published by EWEA, which emphasizes that previously healthy markets such as those of Spain, Italy, and France have recorded very substantial declines compared to 2012: −84%, −65% and −24%. This has resulted in a change in the distribution of the plants which in 2013, unlike in previous years, have focused mainly on just two countries, Germany and the UK, with 46% of new installations. Despite the sharp decline, annual wind power installations in the EU have increased steadily over the past 13 years from 3.2 GW in 2000 to 11 GW in 2013, a compound annual growth rate of over 10% (EWEA 2014; Table 2.7).

With reference to PV energy systems, they play a key role in the transition to a low-carbon energy supply of carbon. This potential has led to the development of efficient PV panels and transformed the area into one of the fastest growing industries. The production of PV cells and modules increased from 46 MW in 1990 to 38.5 GW in 2012. The plants statistically documented all over the world representing almost 100 GW in 2012, putting the EU in a position of advantage with its share of over 69 GW. Actually, the market is shifting towards Asia and America, although the EU remains the main area of installation, the internal market is slowing down for the first time since 2006. In 2012, the EU accounted for just over half the world

Table 2.7 Cumulative wind power installed in Europe by the end of 2013 (MW). (Source: EWEA 2014)

	Installed 2012	End 2012	Installed 2013	End 2013
Austria	296.0	1377.0	308.0	1684.0
Belgium	297.0	1375.0	276.0	1651.0
Bulgaria	158.0	674.0	7.1	681.0
Croatia	48.0	180.0	122.0	302.0
Cyprus	13.0	147.0	0.0	147.0
Czech Republic	44.0	260.0	9.0	269.0
Denmark	220.0	4162.0	657.0	4772.0
Estonia	86.0	269.0	11.0	280.0
Finland	89.0	288.0	162.0	448.0
France	814.0	7623.0	631.0	8254.0
Germany	2297.0	30,989.0	3238.0	33,730.0
Greece	117.0	1749.0	116.0	1865.0
Hungary	0.0	329.0	0.0	329.0
Ireland	121.0	1749.0	288.0	2037.0
Italy	1239.0	8118.0	444.0	8551.0
Latvia	12.0	60.0	2.0	62.0
Lithuania	60.0	263.0	16.0	279.0
Luxembourg	14.0	58.0	0.0	58.0
Malta	0.0	0.0	0.0	0.0
The Netherlands	119.0	2391.0	303.0	2693.0
Poland	880.0	2496.0	894.0	3390.0
Portugal	155.0	4529.0	196.0	4724.0
Romania	923.0	1905.0	695.0	2599.0

Table 2.7 (continued)

	Installed 2012	End 2012	Installed 2013	End 2013
Slovakia	0.0	3.0	0.0	3.0
Slovenia	0.0	0.0	2.0	2.0
Spain	1110.0	22,784.0	175.0	22,959.0
Sweden	846.0	3582.0	724.0	4470.0
UK	2064.0	8649.0	1883.0	10,531.0
Total EU-28	*12,102.0*	*106,454.0*	*11,159.0*	*117,289.0*
Total EU-15	*9879.0*	*99,868.0*	*9402.0*	*108,946.0*
Total EU-13	*2224.0*	*6586.0*	*1757.0*	*8343.0*

market while the previous year had its share of almost three quarters. In 2012, the EU has added to the network 16.5 GW of new capacity for PV, showing a decrease of 25 % compared to 22 GW recorded in 2011.

Over the past 3 years, the growth in the EU was primarily driven by speculative investments that took advantage of the constant difference between the levels of guaranteed remuneration and the very rapid decline in production costs. Some governments have difficulties to overcome this situation and they are trying to mitigate the costs by introducing taxes on electricity production or retroactively changing their own laws. The situation is quite critical so that the European Photovoltaic Industry Association has appealed to the EU for you to take measures against countries that fail to fulfill their commitments on support of RES. It is possible to note the paradox of the EU market where some governments have decided to reduce the volume of installations when it is about to be reached grid parity in their electricity market.

The EU photovoltaic industry is dominated by Germany and Italy, respectively on the first and second place (Table 2.8). Together, the two countries in 2012 accounted for over two thirds of the total number of new solar capacity installed at community level, thanks to respectively 76 and 35 GW. Anyway, there have been a decline in demand in Italy and Germany due to cuts in incentives and uncertainty linked to the trade dispute with China[4].

Lower decrease of installations, compared to 2012, is being recorded even in smaller markets such as Belgium, Denmark, Bulgaria, France, Greece, the Netherlands, Slovenia, and Spain, while the UK, Austria, and Romania in 2013 have grown significantly. In particular, the UK market is the largest market in the first quarter of 2014, especially in the rush to complete the great PV parks before the cuts of the renewable obligation certificate that are British green certificates. After four consecutive quarters of decline, Germany is expected to resume and regain his place as leader of the PV in the second half of 2014 even though the installed capacity in 2014 will be lower than in 2013. To keep alive the German demand, there will be mainly installations in self-consumption; self-consumption will be the spring that will restart the Italian market.

Despite the current difficult economic conditions, the number of new PV markets is increasing. This, together with rising energy prices and the pressure to stabilize the climate, maintains high demand for solar energy systems. The production of electricity by PV modules has already demonstrated that it can be cheaper than conventional electricity currently provided to consumers in many countries.

Concerning biomass, data available in November 2011 on the use of solid biomass in the EU (practically wood for the production of heat or electricity) highlight the increase of the use of biomass during the entire 2010, an increase of 8 % corresponding to an amount passed from 73.4 Mtoe in 2009 to 79.3 Mtoe in 2010. It has

[4] China imposes two kinds of duties: the antidumping and antisubsidy for imports of polysilicon for photovoltaic panels from the USA and South Korea. As part of the Chinese government, the decision should strongly discourage the use of external materials, strengthening domestic production.

Table 2.8 Primary solar photovoltaic production in the EU (GWh). (Source: EurObserv'er and ENEA 2013)

	2010	2011	2012
Belgium	560.0	1170.0	2148.0
Bulgaria	15.0	101.0	814.0
Czech Republic	616.0	2182.0	2149.0
Denmark	6.0	15.0	104.0
Germany	11,729.0	19,599.0	26,380.0
Estonia	0.0	0.0	0.0
Ireland	0.0	0.0	0.0
Greece	158.0	610.0	1694.0
Spain	6425.0	7441.0	8193.0
France	620.0	2078.0	4016.0
Croatia	0.0	0.0	2.0
Italy	1906.0	10,796.0	18,862.0
Cyprus	6.0	12.0	22.0
Latvia	0.0	0.0	0.0
Lithuania	0.0	0.0	2.0
Luxembourg	21.0	26.0	38.0
Hungary	1.0	1.0	8.0
Malta	0.0	8.0	13.0
The Netherlands	60.0	100.0	254.0
Austria	89.0	174.0	337.0
Poland	0.0	0.0	1.0
Portugal	211.0	280.0	393.0
Romania	0.0	1.0	8.0

Table 2.8 (continued)

	2010	2011	2012
Slovenia	13.0	66.0	163.0
Slovakia	17.0	397.0	424.0
Finland	5.0	5.0	5.0
Sweden	9.0	11.0	19.0
UK	40.0	244.0	1188.0
EU (28 countries)	*22,505.0*	*45,317.0*	*67,236.0*

been, 2010, the year in which it has begun to bear fruit the choice of EU countries to exploit more decisively the potential of biomass. In fact, the increase in 2009 was 4% compared to 2008, almost half that recorded in step 2009–2010. This figure shows that also the economic crisis, which started apparent already at that time, has not undermined the efforts made by EU member states to make structural exploitation for energy purposes of biomass. Indeed, the industry has shown a considerable ability to respond to significant increases of the energy demand for heating and for the production of electrical energy, also in a context characterized by great economic difficulty. By looking at the statistics of the past 20 years, this decline in energy production from solid biomass is particularly unusual given that in the EU it has grown steadily since 1990, except for a slight decline in 1999. It is in fact that, in the period 1990–2010, it is more than doubled (39.5 Mtoe produced in 1990).

A stop of the increasing trend in energy production from biomass has been in 2011: primary energy production from solid biomass (wood, wood waste, other biomass plants or animals) in the EU has declined by 2.4 Mtoe reaching 78.8 Mtoe. The gross consumption of primary energy from solid biomass, which includes imports and exports, amounted to 80.8 Mtoe in 2011, a decrease of 3.9% with respect to the previous year. The main reason for this difference is the increase in imports such as wood pellets from Canada and the USA. Final consumption of energy, which is much of the heat[5] consumption (in the residential and industrial sectors), amounted to 58 Mtoe, showing a decrease of 3.1% compared to 2010. The total consumption of heat has been 64.9 Mtoe in 2011, a decrease of 2.4 Mtoe compared to 2010.

It is also interesting the ranking of countries for energy production with primary solid biomass: Germany stands out, followed closely by France, Sweden, and Finland (Table 2.9). Then by looking in more detail the composition of the data, it turns out that when it comes to the use of biomass for the production of heat, the ranking changes, and Sweden, Finland, and Denmark spring at the top.

Eurostat provides biofuels that is any fuel derived from biomass.

Finland, in particular, it is noted that the European country that makes more use of wood for energy purposes has increased the exploitation in a single year (2009–2010) by as much as 18.6%. This pulse is due to the impetus given to the construction of many new cogeneration plants, in line with what happened in the past decades in Finland, period in which there were constructed 50 biomass cogeneration plants and 300 plants for district heating powered mainly by coal and peat. The shift from coal to biomass is also a consequence of a policy of taxation on CO_2 emissions introduced in Finland since 1997. This route is similar to what is making Sweden, where the CO_2 emission has been almost steadily decreasing since, in 1990, it was introduced this kind of taxation.

The Swedish case is particularly significant because despite the carbon tax was in 2011, the highest in the EU (up to touch 115 €/ton CO_2), its effect on the economy was not depressing, but rather gave new impetus to the installation of biomass plants.

[5] Heat production generally comes from cogeneration plants (plants that produce both heat and electricity).

Table 2.9 Primary solid biofuels production in the EU (1000 t of oil equivalent—toe). (Source: Eurostat 2014)

	2010	2011	2012
Belgium	952.1	1104.6	1404.1
Bulgaria	942.5	1030.8	1108.8
Czech Republic	2094.4	2079.7	2152.9
Denmark	1703.2	1498.7	1488.6
Germany	11,762.2	11,054.3	11,810.9
Estonia	957.7	939.0	1012.5
Ireland	190.4	189.9	195.4
Greece	724.9	939.7	1000.3
Spain	4534.1	4818.0	4937.6
France	10,327.1	8895.0	9723.2
Croatia	473.2	638.7	693.6
Italy	3501.2	3953.6	4211.8
Cyprus	5.4	5.0	5.3
Latvia	1732.4	1740.5	1869.5
Lithuania	1002.2	983.3	992.0
Luxembourg	52.2	46.2	47.5
Hungary	1524.2	1429.4	1384.9
Malta	0.6	0.6	0.8
The Netherlands	1083.4	995.4	1094.8
Austria	4673.1	4537.1	4819.9
Poland	5866.2	6350.6	6987.7
Portugal	2582.2	2617.2	2341.3
Romania	3900.0	3475.9	3795.1

Table 2.9 (continued)

	2010	2011	2012
Slovenia	530.8	565.5	560.0
Slovakia	740.4	783.7	800.8
Finland	7741.4	7606.7	7912.9
Sweden	9499.6	8933.6	9563.4
UK	1343.7	1623.0	1809.5
EU (28 countries)	*80,440.9*	*78,835.7*	*83,725.0*

The main driver for growth of biomass is electricity production, regardless of whether or not to happen in cogeneration. The majority of European states, driven from their commitments taken in the national action plans for RES, have created the context for the development of electricity production from biomass by applying incentive mechanisms for the production of green electricity. This strong political will arrive at a favorable moment, when coal plants are aging and require significant investment to make the modernization or the construction of new power plants. This could lead operators to allocate part of their investments in co-firing plants or even 100 % biomass.

This passage of energy from biomass is not easy because the investment choices depend on issues related to combustion technologies, security of supply, the type of fuel used and to their costs, in a time when the demand for biomass is growing more and more. Up to now, the local supply constraints have limited the power supply of biomass power plants under 50 MW. The decision to build plants much more powerful, or to convert coal plants to several hundred mega watt biomass co-firing, will result in an expansion of the range of fuel supply and the related logistic issue. Northern EU countries, which have particularly developed forest industries that generate waste biomass, are ahead of the technical and logistical. They include leading companies in the production market power and cogeneration from biomass.

2.3 EU RES Outlook by Scenario

The latest edition of the future scenarios of the IEA (2014a), the WEO-2014, shows that the pursuit of "climate" requires the development and dissemination of RES much higher than those that current policies and those provided by the scenario of new policies can ensure. The WEO-2014 presents three scenarios, i.e., the new policy scenario, the 450 scenario, and the current policy scenario with a time horizon to 2040. The new policy scenario is based on the continuation of policies that had been legally enacted as of mid-2014 plus plans and commitments that have been announced by countries even if they are not yet implemented. The 450 scenario provides energy pathway consistent with the goal of limiting GHG emissions to an atmospheric concentration that does not exceed 450 ppm CO_2-equivalent, so as to hold the increase in global average temperature below 2 °C. This scenario is considered as a kind of lifeline to preserve the environment from the consequences of the danger of climate change. The current policy scenario takes into account only the policies put in place until the middle of 2014 and it describes a future in which governments do not implement recent commitments nor introduce new policies.

For instance, the 450 scenario promises for the year 2040 a share of RES in the total world demand for primary energy sources of 30 %, compared to 19 % new policies scenario and 15 % current policies scenario, doubling the current share (Table 2.10). This perspective, however, it is possible not only with an increase in the so-called modern RES but also through the progressive replacement of the

Table 2.10 World RES use by type and scenario. (Source: IEA 2014a)

		New policies scenario		Current policies scenario		450 scenario	
	2012	2020	2040	2020	2040	2020	2040
Primary demand (Mtoe)	1802.0	2254.0	3455.0	2223.0	3095.0	2276.0	4658.0
USA	136.0	186.0	346.0	181.0	290.0	188.0	555.0
European Union	198.0	263.0	390.0	255.0	351.0	267.0	504.0
China	316.0	410.0	589.0	397.0	501.0	414.0	842.0
Share of global total primary energy demand	13 %	15 %	19 %	15 %	15 %	16 %	30 %
Electricity generation (TWh)	4807.0	7263.0	13,229.0	7010.0	11,046.0	7329.0	17,973.0
Bioenergy	442.0	764.0	1569.0	740.0	1299.0	768.0	2261.0
Hydro	3672.0	4553.0	6222.0	4458.0	5862.0	4561.0	6943.0
Wind	521.0	1333.0	3345.0	1254.0	2552.0	1376.0	4953.0
Geothermal	70.0	120.0	378.0	113.0	287.0	121.0	557.0
Solar photovoltaic	97.0	449.0	1291	408.0	832.0	459.0	1982
Concentrating solar power	5.0	41.0	357.0	34.0	173.0	42.0	1158
Marina	1.0	3.0	66.0	3.0	41.0	3.0	119.0
Share of total generation	21 %	26 %	33 %	25 %	25 %	27 %	51 %

traditional use of biomass with other forms of energy production with lower environmental impact.

Although the WEO-2014 focuses in particular on the description of the new policies scenario to outline the future of RES, highlighting, for example, the need for a cumulative investment of US$ 7800 billion in the period to 2040 (IEA 2014b), it addresses the specific problems, such as the integration of intermittent RES systems, in particular electrical, the implications for the development of RES markets, and the cost competitiveness of RES compared to fossil fuels, taking in the view 450 scenario a completely different scale. In fact, for the two main intermittent RES, wind and solar, even in the case of new policies scenario it is expected that the installed capacity by 2040 to quadruple in the case of wind farms increases seven times in the case of solar panels, reaching a share of global electricity generation of 6 % for wind power and 3 % for solar PV. This rapid growth is expected to be localized mainly in the emerging global economies, India and China, although the EU, the USA, and Japan should continue to play an important role. In the face of significant developments identified by the new policies scenario, the IEA (2014a) highlights the significant implications in terms of adaptation, new design and investment in energy networks for the exploitation of intermittent sources and in terms of the need for traditional backup capacity, or flexible systems for dispatching energy to cope with the demand for energy that might not be covered by RES.

In the case of 450 scenario, these elements should be addressed in the light of a portion of the RES electricity generation in 2040 that could reach 51 % of the total, with solar and wind power installed capacity equal to almost 50 % of peak demand global: practically a revolution that would require a radical rethinking of energy systems.

Regarding power generation in the new policy scenario, RES accounts for about half of the growth in global electricity generation in 2040, with nonprogrammable sources, wind and solar PVs, which account for 35 % of the increase of RES. In absolute terms, China is showing the greatest increase in electricity production from RES. In some markets, the increasing share of nonprogrammable RES in the generation mix raises issues of primary importance regarding the current structure of the electricity market and its ability to ensure an adequate level of investment and reliability of supply in the long-term. The increase in electricity production from RES means that the share of these sources on the mix of generation worldwide is expected to lie above the 12 % by 2040. For its part, the EU is faced with a major challenge: increase the share of RES, emit less polluting emission, by trying not to raise energy prices through incentives too generous, but even reduce them in a decided to return to being competitive in the world (Table 2.11) It should be emphasized that the EU with this additional effort compared to what has been done so far, it could reach the remarkable share of 47 % of electricity produced from RES.

The EU has adopted the strategy climate and energy package with a goal, among others, of raising the share of EU energy consumption produced from RES to 20 % by 2020. The goal of RES has binding commitments for each country and penalties in the event that are not met. The individual national targets take into account the different starting points, the potential for RES and economic performance. Actually,

Table 2.11 Electricity generation in the EU in the new policy scenario by source. (Source: IEA 2014a)

Electricity generation (TWh)									
	1999	2012	2020	2025	2030	2035	2040	Share 2012 (%)	Share 2040 (%)	CAAGR[a] 2012–2040 (%)
Total generation	2576.0	3260.0	3400.0	3488.0	3563.0	3652.0	3742.0	100.0	100.0	0.5
Coal	1050.0	935.0	788.0	682.0	540.0	402.0	346.0	29.0	9.0	−3.5
Oil	224.0	73.0	37.0	23.0	17.0	16.0	11.0	2.0	0.0	−6.6
Gas	193.0	582.0	583.0	707.0	768.0	838.0	881.0	18.0	24.0	1.5
Nuclear	795.0	882.0	856.0	776.0	790.0	811.0	793.0	27.0	21.0	−0.4
Hydro	289.0	335.0	379.0	391.0	399.0	405.0	410.0	10.0	11.0	0.7
Bioenergy	20.0	170.0	209.0	229.0	248.0	262.0	272.0	5.0	7.0	1.7
Wind	1.0	206.0	398.0	500.0	592.0	672.0	742.0	6.0	20.0	4.7
Geothermal	3.0	6.0	9.0	12.0	15.0	19.0	22.0	0.0	1.0	4.9
Solar PV	0.0	67.0	130.0	150.0	164.0	177.0	189.0	2.0	5.0	3.8
CSP	–	4.0	10.0	14.0	21.0	30.0	38.0	0.0	1.0	8.6
Marine	1.0	0.0	1.0	3.0	8.0	20.0	39.0	0.0	1.0	17.2

[a] Compounded average annual growth rate

it is almost reached at community level the objective relating to the consumption of RES. In fact, in 2011, the energy produced from RES came to 13 %, a figure above 11, 7 % expected for 2011–2012 (Table 2.12).

With reference to electricity, according to the National Renewable Energy National Plans defined by the Directive 2009/28/EC, about a third of the electricity consumption in the EU will come from renewable sources by 2020. In particular, wind energy will be 14.1 % of the electricity consumptions in 2020, hydropower 10.5 %, biomass 6.5 %, PV 2.35 %, concentrated solar power 0.5 %, geothermal energy 0.3 % and ocean energy 0.15 % (European Renewable Energy Council 2011) (Fig. 2.1).

2.4 Conclusions

The analysis carried out in this chapter confirms that the production of energy from RES is a consolidated reality in the EU, able to meet 14 % of EU energy demand in 2013.

It is provided the state of the development of RES in EU member states in terms of installed capacity and energy produced by each technology.

The study shows that the role of RES is expected to strengthen in the coming years if governments decide to increase their efforts against climate change. In fact, if we consider the scenario of the IEA (2014a), which provides for the implementation of the announced policies and the pursuit of international commitments, i.e., new policy scenario, or what they assumed new policies to curb polluting emissions to try to limit the temperature increase to 2 °C, that is the 450 scenario, the share of RES on global demand could grow by as early as 2040 up to 19 and 30 %, respectively. China will drive the change, more than tripling the production of electricity from RES and becoming a leader in the world in 2040, in the new policies scenario, followed by the USA and EU countries as a whole rather than to double its production.

It should be noted, however, that the EU with this additional effort compared to what has been done so far could reach the remarkable share of 47 % of electricity produced from RES, paving the way for a radical change in their energy model with consequences both in terms of infrastructure needs and management of the energy system, both in terms of broader social organization, production, and consumption to be explored. By the way, all these efforts are made by the EU which accounts for 11% of global CO_2 emissions in 2012, which is low compared with other countries such as China (29 % share) or USA (15 % share). What will therefore increase the use of RES overwhelmingly compared to the rest of the world, with all the efforts that result in economic terms? The EU wants to assume the global leadership beyond the costs to bear, or is there a strategic plan of convenience in terms of cost/benefit? This is the crucial question that we will face in the next few chapters of the book.

Table 2.12 EU member states progress towards 2020 targets in RES. (Source: Eurostat 2013)

	2009	2010	2011	2012	2020 target
European Union (27 countries)	11.9	12.5	13.0	14.1	20
Belgium	4.6	5.0	5.2	6.8	13
Bulgaria	12.4	14.4	14.6	16.3	16
Czech Republic	8.5	9.3	9.3	11.2	13
Denmark	20.4	22.6	24.0	26.0	13
Germany	9.9	10.7	11.6	12.4	18
Estonia	23.1	24.7	25.0	25.2	25
Ireland	5.2	5.6	6.6	7.2	16
Greece	8.4	9.7	11.8	15.1	18
Spain	13.0	13.8	13.2	14.3	20
France	12.2	12.7	11.3	13.4	23
Croatia	13.1	14.3	15.4	16.8	20
Italy	9.3	10.6	12.3	13.5	17
Cyprus	5.6	6.0	6.0	6.8	13
Latvia	34.3	32.5	33.5	35.8	40
Lithuania	20.0	19.8	20.2	21.7	23
Luxembourg	2.9	2.9	2.9	3.1	11
Hungary	8.0	8.6	9.1	9.6	14
Malta	0.4	0.4	0.7	1.4	10
The Netherlands	4.1	3.7	4.3	4.5	16
Austria	30.4	30.8	30.8	32.1	34
Poland	8.8	9.3	10.4	11.0	15
Portugal	24.5	24.2	24.5	24.6	31

Table 2.12 (continued)

	2009	2010	2011	2012	2020 target
Romania	22.6	23.2	21.2	22.9	24
Slovenia	18.9	19.2	19.4	20.2	25
Slovakia	9.3	9.0	10.3	10.4	14
Finland	31.2	32.4	32.7	34.3	38
Sweden	48.2	47.2	48.8	51.0	49
The UK	3.0	3.3	3.8	4.2	15

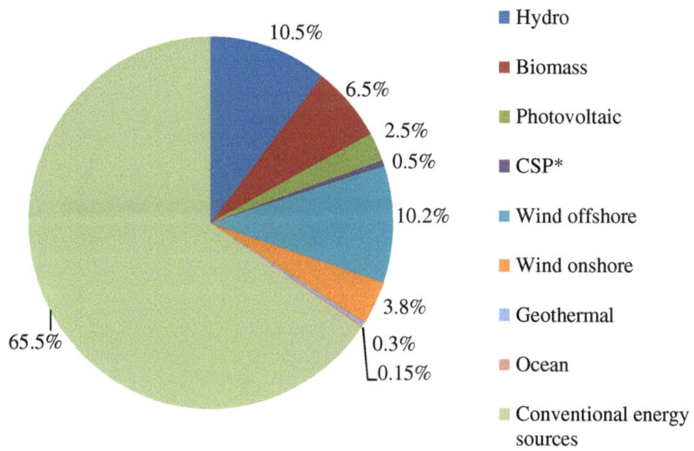

* Concentrated solar power

Fig. 2.1 RES in the electricity mix in the EU in 2020, concentrated solar power. *RES* renewable energy sources. (Source: European Renewable Energy Council 2011)

References

Enea (2014) http://www.enea.it/it/produzione-scientifica/barometri-sulle-fonti-rinnovabili

EurObserv'ER (2014) http://www.eurobserv-er.org. Accessed 27 Jun 2014

Eurostat (2013) http://epp.eurostat.ec.europa.eu/portal/page/portal/statistics/themes. Accessed 1 Jul 2014

Eurostat (2014) http://epp.eurostat.ec.europa.eu/portal/page/portal/statistics/themes. Accessed 3 Jul 2014

European Commission (2001) Directive 2001/77/EC of the European Parliament and of the Council of 27 September 2001 on the promotion of electricity produced from renewable energy sources in the internal electricity market. Off J Eur Union L 283:33–40

European Renewable Energy Council, EREC (2011) Mapping Renewables Energy Path-ways towards 2020, Belgium

European wind energy association (EWEA) (2014) Building a stable future, EWEA Annual Report 2013, Brussels

International Energy Agency (IEA) (2014a) World Energy Outlook 2014, OECD/IEA, Paris

International Energy Agency (IEA) (2014b) World Energy Investment Outlook 2014, OECD/IEA, Paris

Renewables Global Status Report (REN21) (2013) Renewables 2013 Global Status Report. Paris

World Energy Council (2013) http://www.worldenergy.org. Accessed 30 Jun 2014

Chapter 3
Policies for the development of RES in the European Union

In recent years, the renewable energy sources (RES) have grown in importance in the energy system of the EU and in the world; it is expected they will grow more and more in importance in the near future. Since the costs of RES technologies are still higher than the technologies derived from fossil fuels, RES are in need of regulatory intervention. There is still uncertainty on the part of political authorities in relation to the type of policy to better manage the evolution of technologies in the RES market. The support mechanism for RES can be divided into price oriented and market oriented. The focus of this chapter is to analyze the RES support mechanisms among the new EU guidelines on state aid for environmental protection and energy 2014–2020 and in particular the transition from some tools (i.e., feed-in tariffs) to others (i.e., feed-in premium) in order to spur the development of RES. We evaluate the "success" of such instruments in relation to adaptation to the times: while feed-in tariffs can fulfill a variety of policy goals, feed-in premium can reach other objectives. We evaluate the effect of RES support mechanisms in relation to the market structure.

3.1 Learning Curves and RES

To allow an actual success of the European energy market, it is essential that barriers that hinder development be known. In particular, the structural barriers are a major obstacle, since they are related to the actual physical structure of power grids in the individual countries of the EU. Directive 77/2001/EC has intervened to establish an obligation for member states to minimize the obstacles of any kind in order to accelerate the deployment of RES, streamline and accelerate administrative procedures and arrangements to ensure objective, transparent, and nondiscriminatory in the face of unfavorable conditions (European Commission 2001).

Regarding other barriers to the development of RES, they can be divided into four categories:

S. Bigerna et al., *The Sustainability of Renewable Energy in Europe*,
DOI 10.1007/978-3-319-12343-1_3

- Administrative barriers. The first problem inherent to administrative barriers is that in many countries there is a large number of organizations who make decisions about a single market (such as wind power or solar). In this way, it becomes difficult to obtain permits and secondly funding, given that the time is considerably longer. Moreover, the interaction among different organizations might produce substantial inefficiencies. It also adds a problem inherent to local governments: the EU has pushed on the delegation to the local authorities as close as possible to the affected communities regarding the development of RES systems. However, local governments are often poorly informed and lack the expertise to make decisions strategically and economically appropriate.
- Infrastructural barriers. The main problem related to networks is fundamentally linked to their obsolescence, because the European electricity networks are the oldest in the world and are born for the distribution of electricity from conventional power plants. The generation of electricity from RES is different and requires major structural interventions. Moreover, the connection to the network is particularly expensive, as well as getting the permission to do so.
- Social barriers. Although the general opinion of people with regard to RES is substantially positive, there are more and more frequent phenomena of opposition from the inhabitants of the areas in which certain plants should arise. It is the so-called NIMBY ("not in my back yard"), which causes strong opposition from the local population in respect of public works site in their territory. A significant negative contribution to this phenomenon is the fact that very often people do not know the origin of the electricity they consume and even what impacts it has on the environment of reference.
- Economic and financial barriers. The system of incentives and support for RES are still too expensive and recovery times are long. Investors wishing to carry out projects in the field of RES are displaced by the inconsistency of certain policies: in some EU countries, the regulatory framework governing the financial aid from the state is not reliable and varies from year to year without stability. Even lenders have difficulties in facilitating investments to provide loans because of the high risk of the project. Also, there is a strong postponement in the return of those investments that generate substantial cash-flow problems in the short and medium term.

The promotion and encouragement of RES have strong reasons, both economic and geopolitical; in addition to environmental issues relating to obligations arising from international agreements on climate and energy, the other economic–political element of great importance is the security or rather the energy independence of the developed countries. The energy produced from RES is still more expensive than traditional sources. In order for the energy produced from RES to be competitive in the energy market based primarily on energy from fossil fuels, the development of RES must be accompanied by specific incentives and processes of learning by contributing strongly to this development.

The future growth of the economics of energy from RES is shown by the trend of the experience gained; the learning curve relates the cumulative quantitative development of a product to the development of the specific costs. The phenomenon of learning curves was observed already in the 1930s by Wright (1936) for airplane production. The learning curve began to receive attention during the Second World War because the governments were trying to predict the cost and time needed to build ships and aircraft to use during the war (Yelle 1979). Some attempts to use data generated from shipbuilding companies were published by Montgomery (1943).

Arrow (1962) was the first economist to officially note the existence of a learning curve. In "The economic implications of learning by doing" (p. 156), Arrows wrote: "The role of experience in increasing productivity has not gone unobserved though the relation has yet to be absorbed into the main corpus of economic theory". Then he proposed an endogenous theory of changes in knowledge which leads to improvement in production functions over time. Subsequently, many authors have used learning curves to study the diffusion of new technologies and recently a major focus in this regard has been placed at the RES (Gruber the et al. 1999; McDonald and Schrattenholzer 2001).

The learning curve is a tool that can be used both for assessments relating to strategic production competitiveness and to design or reorganize production systems taking into account variations that occur over time as a result of the phenomenon of learning. The basic concept is that the production efficiency of each activity increases continuously to the recurrence of such activities. This basic concept, translated into mathematical models, allows to predict with a reasonable precision, if wisely applied, the variation in the time-dependent variables from learning and progress in which the cost of a product unit, the time required to build it, maintenance hours required per unit volume of production, etc. (Table 3.1).

The learning can be considered as the sum of the factors of discrete and continuous improvements. Discrete factors are triggered by significant events, such as inventions, findings, widespread applications in a very concentrated of fundamentally new technologies that can cause a change virtually instantaneous and easily noticeable in the sector observed. Continuous improvements are events not perceivable individually to a superficial observation and which are attributable to the areas of design, technological/technical and organizational/management.

Table 3.1 Estimated learning rates. (Source: Azevedo et al. 2013)

Technology	Range of learning rates (%)	Time period
Coal	5.6–12	1902–2006
Natural gas	0.65–5.3	1980–1998
Nuclear	0–6	1975–1993
Wind (onshore)	−3 to 32	1980–2010
Solar PV	10–53	1959–2001
Biomass production	12–45	1971–2006
Biopower generation	0–24	1976–2005
Hydropower	0.5–11.4	1980–2001

The main factors that influence the learning curve can be grouped into four categories:

- Positive changes in technology lead to an increase in the learning rate
- Changes in the market such as the entry of new competitors may lead to a variation of the learning rate
- Government policies that stimulate research and development (R&D)
- Furthermore, the learning rates are a set of learning rates of the parts that make up the system. For example, the solar panels are made of modules and other components, each characterized by a learning rate

Generally, learning curves follow the mathematical function through which the measurement of productivity is linked to a parameter related to the measure of productivity itself, the cumulative volume of production, and the learning index. The learning index represents what percentage varies productivity at each percentage variation of the volume of production, that is what is the relationship between the productivity by passing a cumulative volume of production to a multiple.

3.2 Economic Incentives to RES

In order to promote the use of RES, all EU countries have introduced incentives to stimulate the market. However, every reality has its own rigidity and its structure, for which the incentives must be able to adapt and be efficient in often very different contexts. The governments of the EU countries use a variety of instruments to promote the adoption of RES.

There are two basic types of environmental regulation: command and control and economic incentives. These instruments are definitely different in the way they work, but they face the same objective that is stimulating, both in direct and indirect ways, the development of RES so that they will be able to compete with fossil fuel technologies (Held et al. 2006).

Command and control regulation involves an authority collecting the information necessary to decide the action to control and limit pollution; then, polluting firms have to accomplish the steps specified by the authority to reduce polluting emissions and they could resort to the use of RES. Even if the command and control mechanisms give certainty about the level of pollution emitted by firms, it is much costlier than other alternatives such as economic incentives.

Economic incentives give more choice to firms by providing an incentive for the firms to find the best way to reduce pollution, that is, the use of RES and energy efficiency. There are also voluntary approaches: these kinds of strategies are based on consumers' willingness to pay premium rates for RES, such as projects and programs of donation equity. Among economic incentives to improve RES, the important classification criteria are whether policy instruments are price oriented or quantity oriented. With strategies based on price, financial support is given to investment subsidies, soft loans, or tax credits. The financial support is generally given through the feed-in tariff or a fixed premium that governments have to pay for RES produced by the companies entitled. Within the instruments based on the price, the most widely used within the EU is the feed-in tariff scheme under which the supplier or network operators are obliged to buy electricity produced from RES at a higher price compared to the price they pay for energy from fossil fuels, since RES' technologies are younger and therefore production costs are higher. The feed-in tariff incentive allows producers to sell the energy produced at a fixed rate for a specified period of time. In this way, producers have the certainty of the economic return on their investment.

The feed-in tariff is a policy on the supply side of energy that supports the development of RES projects by offering long-term purchase agreements for the sale of RES electricity.

These purchase agreements are generally offered with contracts lasting 10–15 years and are extended for each kilowatt-hour produced. The payment levels offered per kilowatt-hour may be different depending on the type of technology, the size of the project, the quality of resources, and finally the location of the project that best reflects the true costs of the project (Couture et al. 2010).

The authorities can also adjust the levels of payment for installations, reducing in subsequent years, so as to encourage technological improvements. Successful policies related to the feed-in tariff include three key provisions: guaranteed access to the grid, stable purchase contracts and relatively long duration (15–20 years), payment levels based on the costs of RES generation. In some countries such as Germany, the policies also include streamlined administrative procedures to minimize project costs and speed up the rate of development of RES.

In addition, eligibility is extended to all those who are able to invest, including homeowners, businesses, and state and local government agencies, private investors, and nonprofit organizations. One of the most important challenges for policy makers is to determine the right amount of feed-in tariff payments. Taking a look at what is happening in the world, there are a variety of approaches used that reflect diversity in political objectives. The different approaches can be classified into the following categories:

- Based on the actual level of the levelized cost of energy (LCOE) generation from RES. This approach is the most common, it is mainly used in the EU, and it is the method that has proven success in the development of RES
- Based on the value of RES generation to society, expressed in terms of avoided costs
- Offered as an incentive to price fixed without reference to the LCOE for the RES or avoided costs
- Based on the results of an auction process, which can reduce price distortions as it refers to the market. The auction-based mechanism can be applied to different technologies and differentiated based on the size of projects, etc., and it is a variation of the approach based on the cost

The arguments in favor of the policy of feed-in tariff are principally economic in nature. Among these, there is the ability to provide a secure and stable market for investors, stimulate the growth of local industry, and enhance market access for investors to have a quantitative measure of the impact of the generation of RES.

The criticisms made to the feed-in tariff scheme show that a fixed-price system is not compatible with a free market. In this regard, there are often introduced downward adjustments to these rates, for reasons of technological developments which will meet the market. Generation costs in fact are intended to fall with the development of technical improvements.

An alternative approach to the feed-in tariff payments is the feed-in premium, or bonus, above the prevailing market price.

Feed-in premium policies offer a premium above the average spot electricity market price, which distinguishes them from the feed-in tariff payment structure. Feed-in premium payments, either the premium or the total payment, are dependent on the market price for electricity.

The premium payment can be made in terms of two objectives: to represent the environmental costs and therefore the costs incurred to the company, or the best way to approximate the generation costs of RES. In the approach feed-in premium, the electricity generated from RES is sold on the spot market, where an assurance of guaranteed purchase is generally included and thus leads the generation of RES to separate the dynamics of the spot market.

Feed-in premium policies have been offered in some EU countries like Spain, the Czech Republic, Estonia, Slovenia, the Netherlands and Denmark.

Rates of feed-in premium can be differentiated to make payments as close as possible to the costs for each technology type and size of the project. However, since the total revenues of the project depend also from the market price of electricity, it is appropriate to make some considerations. First, the feed-in premium can be constant or sliding. The constant premium policies generally provide a constant, nonvariable, adder on top of the spot market price. In this configuration, the bonus is on top of the market price and does not respond to changes over time and continues to be offered even if the price of electricity rises. In many sources, the premium payments are called "fixed-premium" policies. This is not the same as the fact that with the feed-in tariff that is fixed, because in this latter the total payments are fixed for the duration of the contract, while in the fixed-premium case the total payment varies with the market price. The designers of the feed-in premium policy can also introduce a floor and a cap on the total amount of the premium or the total amount of the payment. If the market price increases, the policy can be answered by sliding the option premium, which can minimize the total cost of this policy by providing a framework for further payments based on costs. There are many ways to determine the payment of the feed-in premium. The first structure exposed refers to the award as a constant, predetermined to be added on the spot market.

The policies of constant feed-in premium create an incentive to generate electricity when demand is high and when market prices are high. The high spot prices, along with the fixed adder on top, tend to encourage the provision when there is most need. The prospect of higher payments can be seen as a compensation for the higher risk market. However, since a constant is added premium to the spot price market, the additional payment is agnostic to spot market prices. All this can result in average levels of higher payments when electricity prices increase, which then increases the pressure on the total costs of the policy.

This is confirmed in an analysis of the policies of feed-in premium constant, where the average level of payments is about 1–3 cents/kWh higher than those with feed-in tariff policies (Couture et al. 2010). In addition, the pattern of feed-in premium does not consider that electricity prices may decline sharply, so it is possible that the revenues are insufficient to cover the costs. This uncertainty about future revenues creates an additional risk for investors, who are likely to increase the required equity returns and potentially the debt interest rates, which increase the marginal costs of RES deployment.

In response to potential problems related to the approach of the premium constant, some jurisdictions allow the feed-in premium to vary inversely with respect to the market price in oder to stabilize the profits. As for the feed-in tariff policies, it is possible to differentiate the premium to reflect as best as possible the generation costs of RES. The premium differentiation is important as differentiating tariffs in fixed-price options. It allows that the expected amount of payment approximates the actual levelized costs of technology development, while still retaining the market orientation. By differentiating the premium amount, countries using premium policies can differentiate the payment level awarded per kilowatt-hour with greater accuracy, although it is unlikely that the actual payment levels are as close to flush with the cost of generation as it happens with the fixed-price option.

With regard to strategies based on the quantity, the desired level of energy produced from RES or market penetration is defined by governments. Among quantity strategies, the most important tools are the tender system and the tradable certificates system. In the tender system, notices are carried out on a regular basis and for a defined quantity, and the contract is given to the bidder that offers the lowest price. The winners of the tenders get a fixed price per kilowatt-hour for the period of the contract, and the contract itself offers the winner a set of favorable conditions for investments. In the system of tradable certificates, companies that produce energy are required to provide or acquire a certain percentage of electricity from RES. Subsequently, at the date set in the agreement, they must submit the number of certificates to demonstrate compliance.

The companies involved in the tradable certificates system can obtain a certificate from their generation of RES electricity; they may also buy electricity from RES and associated certificates from another generator, or they can buy the certificates that have been marketed independently by their production of energy.

Economic incentives for RES vary between EU member states (Table 3.2).

For instance, in Germany, the main program of support for electricity is represented by an incentive based on price, i.e., the feed-in tariff scheme. The law states that tariffs do not depend on the market price of energy but are defined in the law and that the rates are different for wind energy, biomass, photovoltaic, etc. Moreover, these rates decreased over time in order to take account of the technological learning curves and then improvements. The UK has been the first European

Table 3.2 RES support mechanisms by main technology and country in 2011. (Source: Council of European Energy Regulators 2013)

Member state	Hydro	Wind	Biomass and waste	Biogas	Photovoltaic	Geothermal
Austria	Investment grants, feed-in tariff	Feed-in tariff	Feed-in tariff	Feed-in tariff	Investment grants, feed-in tariff	Feed-in tariff
Belgium	Green certificates with guaranteed minimum price	Green certificates with guaranteed minimum price	Green certificates with guaranteed minimum price	Green certificates with guaranteed minimum price	Green certificates with guaranteed minimum price	Green certificates with guaranteed minimum price
Czech Republic	Feed-in tariff, feed-in premium	Feed-in tariff, feed-in premium	Feed-in tariff, feed-in premium	Feed-in tariff, feed-in premium	Feed-in tariff, feed-in premium	Feed-in tariff, feed-in premium
Estonia	Feed-in premium	Feed-in premium	Feed-in premium	Feed-in premium		
Finland	Excise tax return	Excise tax return	Excise tax return	Excise tax return		
France	Feed-in tariff	Feed-in tariff, call for tenders	Feed-in tariff, call for tenders	Feed-in tariff	Feed-in tariff, call for tenders	Feed-in tariff
Germany	Feed-in tariff, direct marketing, feed-in premium	Feed-in tariff, direct marketing, feed-in premium	Feed-in tariff, direct marketing, feed-in premium		Feed-in tariff, direct marketing, feed-in premium	Feed-in tariff, direct marketing, feed-in premium
Hungary	Feed-in tariff	Feed-in tariff	Feed-in tariff	Feed-in tariff	Feed-in tariff	Feed-in tariff
Italy	Green certificates, feed-in tariffs	Green certificates, feed-in tariffs	Green certificates, feed-in tariffs	Green certificates, feed-in tariffs	Feed-in premium	Green certificates, feed-in tariffs
Lithuania	Feed-in tariff	Feed-in tariff	Feed-in tariff	Feed-in tariff	Feed-in tariff	
Luxem-bourg	Feed-in tariff, feed-in premium	Feed-in tariff, feed-in premium		Feed-in tariff, feed-in premium	Feed-in tariff, feed-in premium	
Nether-lands	Feed-in premium	Feed-in premium	Feed-in premium		Feed-in premium	
Norway		Investment grants				
Portugal	Feed-in tariff	Feed-in tariff, tender-ing process	Feed-in tariff, ten-dering process	Feed-in tariff	Feed-in tariff	

Table 3.2 (continued)

Member state	Hydro	Wind	Biomass and waste	Biogas	Photovoltaic	Geothermal
Romania	Green certificates	Green certificates	Green certificates	Green certificates	Green certificates	
Slovenia	Feed-in tariff	Feed-in tariff	Feed-in tariff	Feed-in tariff	Feed-in tariff	
Spain	Feed-in tariff or feed-in premium (optional)	Feed-in tariff or feed-in premium (optional)	Feed-in tariff or feed-in premium (optional)	Feed-in tariff or feed-in premium (optional)	Feed-in tariff (PV) and feed-in tariff or feed-in premium (CSP)	
UK	Green certificates, feed-in tariff	Green certificates, feed-in tariff	Green certificates		Green certificates, feed-in tariff	

RES renewable energy sources, *CSP* concentrating solar power , *PV* photovoltaic

country to pursue liberalization of the electricity market. In the UK, the production of energy from RES is supported by strategies based on quantity. Since 1999, the system has adopted a quota system with tradable green certificates bond. The objective is that obligation increases over years, and power companies that do not comply with the obligation have to pay penalties.

However, the prevalent mechanism in the EU is the feed-in tariff scheme and it has also provided good results; green certificates, on the contrary, have had their heyday in the past and then interest has decreased because of the problems linked to the higher uncertainty that they leave in the tariff market.

RES technologies like wind and solar have been heavily subsidized since the EU set a target in 2008 of sourcing 20% of energy from RES by 2020. Critics say this has led to market distortions in the energy market which have raised prices, most notably in Germany. The difference between the market price for electricity and the higher fixed price for RES is passed on to consumers, whose bills have been rising for years. Household electricity prices have risen by 4% a year between 2008 and 2012 as a result of levies and taxes linked to RES, according to the European Commission.

From 1970 to 2012, RES power has received 3.4 c/kWh of financial support on the average. RES electricity is largely paid for with feed-in tariffs per kilowatt-hour of power. The specific rates paid depend on the kind of technology used: in 2012, the average feed-in tariff was 8.8 c/kWh for wind power onshore, 8.5 c/kWh for hydropower, and 36.5 c/kWh for photovoltaic; the figures include older installations that still receive higher rates than are paid today, especially in the case of photovoltaic (Table 3.3).

The implementation of new policies to promote RES has a considerable impact on the amount of investments made in this sector.

In order to reach the RES target, an investment of € 443 billion in RES is needed over the period 2001–2020, of which € 303 billion from 2010 to 2020. The EU countries and unit support levels on final electricity consumed varied between 0.12 and 20.61 €/MWh, while the average level of support was found to be approximately 7 €/MWh in 2010 (CEER 2013).

Currently, cuts in RES incentives have been widespread in the EU in recent years, and particularly since 2012 as governments realized that subsidies had been incorrectly set. Germany and the UK both brought in sudden, unscheduled cuts to solar support last year, while Spain, by far the world's largest photovoltaic market back in 2007 and 2008, suspended all incentives for new projects at the start of 2012. Italy, the largest solar market in 2011, set an overall 6.7 billion € cap on subsidies for that technology in 2012. More damaging to investor confidence than new caps or sudden reductions are retroactive cuts in support, which reduce the revenues of RES projects that are already operating. Spain was the first to announce retroactive cuts at the end of 2010; in its case, as part of an attempt to address a multibillion-euro energy system deficit that was technically a government liability. It has since been followed by the Czech Republic, Greece, and Bulgaria. In some cases, these curbs on existing wind and solar plants have taken the form of monthly basis and the UK followed in 2012. The UK also has a levy control framework

Table 3.3 RES support levels by main technology and country in 2011 (MWh). (Source: Council of European Energy Regulators 2013)

EU member state	Hydro	Wind	Biomass	Biogas and waste	Photovoltaic	Geothermal	Total
Austria	1.13	21.55	81.12	98.2	263.64		46.49
Belgium	45.17	94.58	96.57		407.42		142.04
Czech Republic	57.08	63.56	55.06	107.5	432.33		196.32
Estonia	51.61	53.68	53.68	56.25			53.66
Finland	4.2	11.97	6.74	4.2			6.93
France	13.17	33.04	54.85	41.45	477.22		116
Germany	48.66	45.43	143.74	25.97	353.82	157.59	130.77
Hungary	71.78	111.48	112.97	108.77			107.33
Italy	70.3	69	119.904		367.2	80	153.69
Luxembourg	79.33	36.38		70.46	543.43		138.21
The Netherlands	103.93	68.47	75.11	41.33	385.88		70.89
Norway		11.27					11.27
Portugal	40.54	42.68	49.16	39.5141	291.78		47.03
Romania	59.81	65.17	63.7742		78.74		64.39
Slovenia	23.47	95.38	87.24	126.76	343.07		81.05
Spain	39.02	40.94	75.11	31.26	356.76		84.8
Sweden							21.47
UK	64.81	72.71	58.48	62.8	290.37		59.92
Minimum support	1.13	11.27	6.74	4.2	78.74	80	6.93
Maximum support	103.93	111.48	143.74	126.76	543.43	157.59	196.32

RES renewable energy sources, *EU* European Union

that limits how much can be added to consumer energy bills to fund clean energy projects. The aim is to help to reassure investors that costs will not spiral towards sudden or retroactive policy changes.

With reference R&D investments, under the 2020 Climate and Energy Package, the Strategic Energy Technology Plan has increased R&D investments across the EU from € 3.2 to 5.4 billion per year and is progressing towards a single, integrated roadmap to guide future investments. For the 2014–2020 period, the EU is ramping up investment in energy- and climate-related R&D and under Horizon 2020, the new Union research and innovation program, close to € 6 billion will be dedicated to energy efficiency and to secure clean and low carbon technologies and to smart cities and communities.

It should be noted that conventional energy has received subsidies for production or consumption of fossil fuels for decades.

Fossil fuels and nuclear energy have been developed under protected, monopolistic market structures. Between 1970 and 2010, in Germany for instance, coal received € 222 energy billions, nuclear received € 186 billion, and RES received € 28 billion. Looking more specifically at R&D spending: nuclear energy received more than € 12 billion since 1984 out of the EU's research budget, while nonnuclear energy (fossil fuels, carbon capture and storage (CCS) and all RES electricity, heating, and transport technologies) received € 6.5 billion (Fig. 3.1).

The EU member states have a common approach to the taxation of energy products that was launched with the Energy Taxation Directive 2003/96/EC (European Commission 2003), which was adopted in 2003 and which was revised in 2011 in order to promote energy efficiency and consumption of more environmentally friendly products and to avoid distortions of competition in the single market. Agriculture often benefits from reduced tax rates, as well as navigation and rail

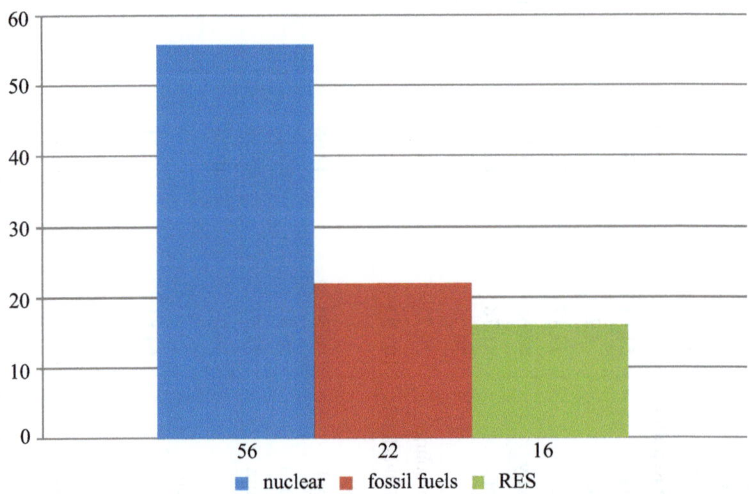

Fig. 3.1 R&D spending in CEM countries from 2000 to 2010 (billion dollars). *R&D* research and development, *CEM* Clean Energy Ministerial. (Source: EREC Factsheet 2014)

transport; domestic aviation is instead often exempted. For instance, in Italy there are existing exemptions for fuel used for certain types of carriage of goods by road and public transport, or for fossil fuels used by consumers living in disadvantaged areas. In some northern European countries, the use of fossil fuels for heating is exempted or taxed at a reduced rate. It is evident that a distributive principle has inspired the support measures mentioned above, aimed at sectors, consumers, or uses for which governments have found it desirable to reserve a leniency. Some EU member states have also reduced rates or exemptions in force for the energy-intensive sectors, in order to avoid impact on competitiveness or reduce the possibility of overlap of policy instruments, as these areas are already involved in the EU emissions trading system (ETS).

Market Structure A classic problem in economics is whether a more competitive market structure has a faster rate of innovation. In this context, it is appropriate to assess the innovations related to market power and subsidies system that supports the RES. The latter compete with the industry of fossil fuels that can be seen as a Cournot oligopoly, at least in the electricity markets of Organisation for Economic Co-operation and Development (OECD) countries that are dominated by a small number of large firms. Regarding the RES, it is appropriate to divide the industry in an upstream sector and a downstream sector. The upstream sector of RES or the producers of equipment and the downstream sector refers to RES producers who buy the equipment upstream sector and sell electricity to consumers (Reichenbach and Requate 2012). The market structure in the market of RES equipment is not so predictable. For example, the ten market leaders in the wind turbine industry account for 80 % of the world market in 2011. They are Vestas (12.7 %, Denmark), Sinovel (9 %, China), Goldwind (8.5 %, China), Gamesa (8 %, Spain), Enercon (7.8 %, Germany), GE Wind (7.7 %, the USA), Suzlon (7.6 %, India), Guodian United Power (7.4 %, China), Siemens (6.3 %, Germany), and Ming Yang Wind Power (3.6 %, China).

Most of the literature that considers that the points for which competitive markets to promote innovation are:

- In competitive markets, there are many firms, so they must work hard to attract customers through innovation and thus developing better products and services, and also offering a greater variety of choice; for example, the Internet is a competitive industry but companies offer different services and products.
- If firms are inefficient, they can easily run into losses, so there is an incentive to innovate and try new ideas and products to attract customers; an example is the telecommunications and in particular from mobile phones that have various applications and accessories and in a monopoly there would be less incentive to innovate because they do not have to compete with anyone.

Points at the expense of competitive markets with respect to innovation are:

- If the markets are too competitive, profits will be low and therefore companies do not spend some of their profits on R&D and therefore innovation will be low.

- Competition implies that there is an exchange of information between firms, with the risk that new products can be copied less than a patent that protects the firm that innovates, but in this case there is no more competition.

In general, it can be said that the competition might promote innovation but in reality this is not always true because a lot of innovation came from monopolies.

The Lee and Wilde (1980) model is a one-shot noncooperative game in which there are *n* identical firms that invest in R&D in order to innovate first. The firm that innovates first wins an exogenously given price. Firms that have not arrived first instead get nothing and even lose what they have invested, that is the cost in R&D. The authors assume that there is an infinite and perfect protection of the patent. The probability of success of the representative firm is a function of its spending on R&D; this expenditure is a flow cost that the representative firm pays until a player earns a win. The authors show that an increase in the number of firms increases the equilibrium relative to the individual effort in R&D. Furthermore, assuming that the social benefit of innovation is equal to the price obtained by the winner, Lee and Wilde show that the equilibrium industry firms invest more than what is socially optimal. These results clearly depend on the assumptions of the model, which means the price is exogenous and independent of the number of firms, those who lose they get nothing, and so on.

Delbono and Denicolo (1991) follow the basic model of Lee and Wilde but do not reach the same conclusions. They show that an increase in the number of firms may lead to a reduction in the equilibrium R&D effort of each firm and in the equilibrium total effort; in equilibrium, there may be underinvestment with respect to the social optimum.

In general, theories of industrial organization believe that innovation can reduce the competition while empirical studies find that innovation increases.

Aghion et al. (2005) analyze the relationship between product market competition and innovation, finding a strong evidence of an inverted-U relationship. The authors assume that both the current technological leaders and their followers in any industry can innovate, and innovations of leaders and followers take place step by step. Incentives for innovation depend on the difference between post-innovation and pre-innovation rents of incumbent firms. In this case, greater competition can stimulate innovation and growth, because it can reduce a firm's pre-innovation rents by more than it reduces its post-innovation rents. This means that competition may increase the incremental profits from innovation and thus encourage investment in R&D aimed to gain market share. This reasoning should be especially true in areas where the incumbent firms operate with fairly similar levels of technology; in these areas, pre-innovation rents should divert product competition. On the other hand, in areas where innovations are made by laggard firms with already low initial profits, product competition may affect the post-innovation rents, and therefore can dominate the effects on Schumpeterian competition. In conclusion, the authors find that competition and innovation are characterized by an inverted-U shape, with the industry growing distributed on both sections that decrease the U-shape.

3.3 Costs of RES in the EU

To compare the cost of the energy produced from RES and fossil fuels, the value of the energy rather than private costs should be considered. The kilowatt-hour produced from RES has the same value and quality of that produced by fossil fuels; the latter has hidden costs that are paid, often unconsciously, by the whole of society. They should be given the environmental and social damage resulting from the generation of energy from conventional forms, although it is difficult to give an economic value to such damages.

The LCOE provides an estimate of the cost of electricity generation from a particular technology in US$/MWh. It includes the cash outflows for the initial investment (either via equity or from loans), interest payments in the case of financing, operating costs, and maintenance costs (for biomass plants, there are cost of the raw materials purchased). The LCOE is declining for most of the RES technologies as wind, solar photovoltaic, and some biomass (Table 3.4).

The cost of photovoltaic energy is higher than that of other RES; however, in the next few years it is estimated that a reduction in the cost of photovoltaic energy production will be much higher than more competitive RES. The research in the photovoltaic field has developed and continues to develop, essentially in two directions: the first is to use materials other than silicon and more complex devices to increase the conversion efficiency of the photovoltaic cells, i.e., the ratio between the generated electric power and solar incident; the second is to lower the cost of the cell in order to make a more competitive cost of electricity produced. The investment cost and manufacturability of a photovoltaic system are the only factors that determine the cost of photovoltaic electricity and are, in turn, strongly dependent on the characteristics of the material used and the processes used in the manufacture of the device or, in other terms, of conversion efficiency and the production cost of the device itself.

Photovoltaic energy is produced by an expensive technology, noncompetitive with traditional forms of electricity production from fossil fuels if not suitably subsidized.

Anyway, over the past 40 years, the price of some modules for photovoltaic panels has decreased due to the learning curve with an average rate of about 40 %. This development was driven by both technological developments and market conditions. It is interesting to note that between 2004 and 2008, the price of photovoltaic modules has remained constant at about US$ 4 per watt peak. This happened despite the fact that in the same period there were improvements in technology and the companies had increased their production. The reason is that the expansion of the photovoltaic market above all in Germany and in Spain, where the supports feed-in tariffs have attracted investors, was accompanied by a reduction in the availability of polysilicon over the same period, which reduced the production of silicon and therefore influenced the pricing competition. The temporary reduction stocks of silicon has led to an increase in investment in thin-film capacities in the same time frame, so that the market share for thin-film modules is increased up to 2009 where it reached almost 40 %. Between 2008 and 2012, there was a massive fall in the

Table 3.4 Estimates of costs of electricity generation (US$/MWh). (Source: Hayward et al. 2011)

	Wind	Solar thermal	Solar PV	Biomass	Geothermal	Wave	Natural gas	Coal	Coal with CCS	Nuclear
2011	63	181	270	75	83	105	47	71	121	76
2030	56	139	211	70	78	103	39	85	96	74

PV photovoltaic, *CCS* carbon capture and storage

price of the modules, to nearly 80 %, resulting in a series of financial problems to the companies that had invested in the sector. The photovoltaic system prices have followed the reduction in module prices but at a lesser rate; this is clear when one considers that the share prices of the modules in the system takes 50 % of the price.

The price of photovoltaic varies between countries. The reasons for these differences lie in the fact that there are different legal requirements for permits, license, and connecting to the network, which then have an impact on photovoltaic manufacturers.

In the years 2012 and 2013, the cost of PV modules has fallen by about 60 %. At the beginning of 2012, the thin-film module prices have fallen below US$ 1 per watt, with prices ranging between US$ 0.84 and 0.93 per watt. The price of crystalline silicon modules is slightly higher and varies between US$ 1.02 and 1.24 per watt.

Germany has recorded in 2010 the lowest level of cost for the photovoltaic system in the small-scale residential market (lower than 5 kW), with an average price for silicon of approximately US$ 4 per watt. Grid parity with the prices of residential electricity market should soon be reached with regard to photovoltaic in areas with good solar radiation, with a few exceptions. The photovoltaic also has the advantage that once it is installed it is very productive and therefore fails to meet the needs of the electricity sector like no other RES is capable of doing.

Wind energy is one of the most competitive technologies among RES and in countries characterized by the massive presence of wind, it is even able to compete with fossil fuels based technologies. In Europe, the most competitive regions for the presence of wind are the north and west coasts of Scandinavia, the UK, and some parts of the Mediterranean.

The assumptions considered to calculate the cost of energy produced by on-land wind power are essentially the following: calculations are related to new land-based, the turbines is a medium size, the lifetime of a turbine is set at 20 years, and the discount rate is assumed to range from 5 to 10 % annum.

After a first period in which the price of the turbines was higher due to an increase in the price of commodities and an excess of demand over supply, the price of the turbines has decreased recently, thanks to the low prices on the market of wind turbines produced in emerging countries. On average, the LCOE of new wind farms in 2010 was around US$ 0.1 per kilowatt-hour, with a cost of capital of 10 %.

The rapid growth of PV in Europe can be demonstrated considering projections to 2010 that had been made by the European Commission in the White Paper in 1997 (COM (1997), p. 599 final), according to which it would have to reach a production amounting to 40 GW. This value was equal to 16 times the installed capacity in 1995, but this goal was reached in 2005 and in 2009 the installed capacity exceeded 72 GW. The majority of wind power is installed onshore, but there are recent offshore sites that have been installed in Denmark, Ireland, Sweden, Germany, the Netherlands, and the UK, so as to reach a production of 1500 MW in 2009. The offshore wind farm is becoming increasingly attractive to many countries because the wind intensity is much higher and the environmental impacts are lower compared to wind power onshore.

With regard to biomass, energy production is quite competitive in countries where agricultural and forestry waste are available at low cost. Energy from biomass varies from mature technologies to new technologies that have not yet been developed on a large scale. This diversity is reflected in the total cost of installing a biomass power plant: in 2010, it varied between US$ 1800 and 6800 per kilowatt. If biomass is used in conjunction with fossil fuels, the costs go down considerably. In 2010, operations and maintenance costs contribute significantly to account for 20 % of LCOE for biomass power plants.

Hydropower is a mature technology and its LCOE is generally very low. The LCOE for large hydro is around US$ 0.04 per kilowatt-hour and it can reach US$ 0.20 per kilowatt-hour in the case in which the dams are also used for other purposes. The cost is so little that it makes hydropower technology more competitive among existing technologies to produce electricity. Small hydro-costs are slightly higher but still competitive.

3.4 Feed-In Tariff Versus Feed-In Premium in the EU Guidelines on State Aid for Environmental Protection and Energy 2014–2020

Currently, the European Commission, through the adoption of specific guidelines on state aid for environmental protection and energy 2014–2020 (IP/14/400) has set new posts reserved for state aid to RES. The new guidelines are going to replace the 2008 State aid Guidelines on Environmental Protection from 1 July 2014.

The aim of the measures contained in the new guidelines is to reduce the distortions that have characterized the incentives in recent years, in particular with respect to the weight that the incentives have had on bills paid by European consumers.

The European Commission has also required member countries to reduce taxes on energy companies so that they might recover competitiveness at the international level.

These, in fact, have judged the punitive measure for the future deployment of RES, as well as recognize by the EU, an excessive interference in national energy policy and a will not fully supported. The guidelines have been designed to gradually replace the incentives for RES with a competitive process based on public auctions. It is, indeed, a setting that finds supporters in our country, especially in reference to the weight, in recent times, the incentives for RES have become part of the electricity bill, a matter of concern to consumers, especially in times of economic crisis.

By analyzing the text of the guidelines, particularly with regard to changes that affect the national promotion of RES, for example, in 2016 the plants above 500 kW (over 3 MW for wind power) will be subject to balance responsibility and therefore cannot be further encouraged by incentives in cases where the power generation takes place in a context characterized by negative prices of megawatt-hour, as it happens in some electricity markets in the EU.

In this regard, the pilot project regarding procedures for auction for a small portion of the new powers are expected to begin in 2015 and from 2017 all new installations will have incentives through competitive mechanisms, although for the European Commission provided for the possibility of a few exceptions to the member states to ensure a certain degree of flexibility in the management of domestic typicalness over the adoption of these new mechanisms.

However, auctions are excluded from the lower power plants (less than 3 MW for wind power and 500 kW for other RES).

These, in fact, will continue to benefit from all types of support, including feed-in tariffs. Other exemptions from the revision of the system of incentives are provided for systems with a capacity of less than 6 MW for wind power and 1 MW for the other RES, provided they are in the initial stages of development (European Commission, 2013, p. 440 final).

Treatment for energy-intensive industries is different with respect to payment of the exemptions in support of RES: in fact, these will be evaluated in consultation with the new retroactive guidelines. The measure also expresses the ability to pay, that is, the contribution made by some countries at the thermal power plants to compensate for the reduction in revenues due to the RES. This is an incentive assigned for the replacement of the production of energy from these plants, in the hours when RES do not produce, as part of an electrical system that has not yet adapted to the management of intermittent sources such as the RES.

In this sense, the decision of the European Commission states that member states, in the future, before considering mechanisms that compensate for the intermittence of RES, must try to remove the distortions that do not allow the market to provide the right incentives to invest in new capacity. Only after this, compensation can be introduced and it is not supposed to restrict competition in the internal energy market. According to the decision of European Commission, member countries will only have one year from the publication of the guidelines in the official gazette of the EU to adapt its system of incentives to the new rules European Commission 2014). It remains to be seen whether some countries will exercise their right to self-determination of the mix of energy policies. Some countries, such as France, Germany, UK and Italy, have sent a protest letter to the Commission during the consultation prior to enactment of the guidelines.

Regarding the agricultural sector, the new guidelines should not have a very strong impact by virtue of the fact that agro-livestock investments are characterized by small plants. This does not mean that the Commission's approach does not raise debates on the costs of RES, threatening to crush the system without making the right distinctions in terms of cost and environmental and energy sustainability, including the speculative interventions and positive that can come within the territory.

The new guidelines provide a framework for designing more efficient public support measures that reflect market conditions, in a gradual and pragmatic way.

The new guidelines provide a framework to outline measures to support the public more efficient that reflect the market conditions in a gradual and pragmatic way. The guidelines aim to outline the state aid that contribute to achieving the 2020 objectives for sustainability and energy security, while ensuring the society contained spending and a reduction in distortion with respect to competition or fragmentation of the single market. The guidelines will be in force until the end of 2020. With these new rules, subsidies for RES are replaced by market-based mechanisms for all generators of electricity by 2017, following a pilot phase starting in 2015 and feed-in tariffs will be replaced by the feed-in premiums that allow RES to capture the signals of the market, while the energy infrastructure and cross-border schemes will be protected in some way, and 68 energy-intensive sectors will be selected for grants.

According to the feed-in premium, the price of RES is composed of two factors: the value of the electricity market is exposed to fluctuations in supply and demand, and a premium fixed by the state. The feed-in premium scheme is therefore going to replace the feed-in tariff scheme and is quite controversial weigh the two schemas to understand what is actually the most effective. Looking specifically at the pros and cons of the two instruments, such as the feed-in tariff, eliminates the risk of price. Detailed analyses of the average levels of payment for a number of feed-in tariff policies in the EU have shown that the policies of fixed-price feed-in tariff have had, on average, a higher degree of cost efficiency with respect to the feed-in premium schemes; this implies lower payments per kilowatt-hour for electricity generated from RES.

The stability of long-term payments with a fixed price leads to lower risks for both the RES project developers and investors and therefore it is more likely that there is a reduction in costs financing. Understood in this way, the feed-in tariff policies are a way to remove price risk, which can decrease costs per kilowatt-hour of RES deployment. In addition, the feed-in tariffs approximate better the actual costs of RES generation if the feed-in tariff rates are set appropriately. This payment structure based on cost is more likely to encourage investment in RES projects, because it actually captures the true costs of the project. The feed-in tariffs are typically accompanied by a guarantee of purchase which further reduces the risk of the market. The assurance that a counterparty will purchase the electricity significantly reduces the risks by providing the certainty of revenues. The feed-in tariffs can actually act as a hedge against energy and electricity price volatility by introducing fixed-price supply into the electricity supply mix.

This effect may help to reduce the wholesale price of electricity at the time when the cost of offering the RES is less than the marginal cost of the offer conventional. So, having a portfolio of electricity generation that includes a fixed price for the RES, the court may protect ratepayers through reduced exposure to the volatility of energy prices. This is very important in markets where a significant proportion of electrical generation comes from natural gas. The main disadvantage of policy feed-in tariff is that it does not adapt to the market price of electricity. These prices are blocked, often in long-term contracts, and typically do not create an incentive for project operators to adjust their production in relation to demand. It was also argued that the feed-in tariff policies, which offer contracts to sell electricity in the long term at fixed prices, may create distortions in the wholesale and retail electricity markets.

Finally, unless the rates are not differentiated by the location of the projects, it is possible that fixed pricing policies do not create incentives to develop electric resources where there is actually needed. In particular, the policies of fixed-price feed-in tariffs that provide interconnection network cannot provide incentives for development in areas prone to congestion, where the spot market prices tend to be higher, or alternatively, where the marginal value of the new generation is higher.

The feed-in premium policies have many advantages that cannot occur in the approach instead of a fixed price. They are more market oriented than the fixed-price schemes because the premium payments depend on the prevailing price of electricity. Consequently, this scheme can create incentives to produce electricity when demand is high and encourages the installation of new generation in areas with higher average market prices for the pricing structures related to localization. In addition, this market orientation can lead to a more efficient management of the network and better provision of related services. Electricity generators are requested on the spot market with the policies of feed-in premium, so that the generators of RES compete with each other and with conventional generators. So, these premises lead us to think that the feed-in premium scheme will encourage more competition among electricity producers. Although the feed-in premium scheme involves some disadvantages, the policies feed-in premium have demonstrated a lower degree of cost efficiency compared to the feed-in tariffs, which involves payments higher average per kilowatt-hour. This is primarily due to the fact that the premium-price option requires higher risks because revenues are less certain. The highest risks require higher returns and this translates into higher costs per kilowatt-hour for the community as a whole, if it is intended however to achieve a certain level of the RES.

The feed-in premium policies generally do not include a guarantee of purchase. Those who participate in the option premium sell their electricity on the spot market and receive the corresponding market price, with the added feed-in premium. Investors perceive the absence of a guarantee of purchase as an additional risk premium in the option, and this mechanism leads to demand higher returns on investment. As projects related to wind and solar cannot influence the time that offer electricity network, it is likely that these technologies can benefit to a lesser extent the option premium. So, while the feed-in premium scheme can provide useful incentives for projects of hydropower, solar thermal electric, biogas, biomass, and solar energy can be penalized because they cannot adapt their supply with the trend of market prices. However, the adoption of feed-in premium scheme should be seen as an adaptation to the times.

There are pros and cons of these tools are changing over time. The EU moved as a pendulum, switching from feed-in premium to feed-in tariff in the past and now it returns to the feed-in premium.

Since 2016, competitive bidding processes will be the rule, and then the power generators will have to sell electricity on the market taking into account the short term. In particular, in 2015–2016, EU member states will begin to implement competitive bidding procedures for small shares of their new capacity from RES. From 2017 onwards, EU member states shall set up tenders to grant support to all new installations. EU member states will be obliged to use the premiums above the market price, as support tools for integrating RES in the market.

3.5 Conclusions

The introduction of subsidy programs for RES technologies and the setting of long-term targets in the energy policy have created a favorable climate for the development of RES. The support mechanisms are designed to encourage learning by doing and innovation, by improving the competitiveness of RES compared to fossil fuels. Drawing from experience in the EU energy system, this chapter offers a critical overview of the existing policies supports for RES, with a focus to price-oriented mechanisms. Regardless to the preferred support mechanism, RES supports have broad and multiple effects and thus they are required to be contextualized with respect to the time period.

Currently, EU countries experience a growing need for RES, for climate change commitments, and also as an integral part of the package of recovery from the economic crisis, with the aim of stimulating green industries. This need has highlighted the importance of energy tariff design as a critical component of RES.

References

Aghion P, Bloom N, Blundell R, Griffith R, Howitt P (2005) Competition and innovation: an inverted-U relationship. Q J Econ 120(2):701–728

Arrow KJ (1962) The economic implications of learning by doing. Rev Econ Studies 29(3):155–173

Azevedo I, Jaramillio P, Rubina E, Yehb S (2013) Technology learning curves and the future cost of electric power generation technology. Paper presented at the 18th Annual Energy and Climate Change Research Seminar, Electric Power Research Institute, Washington, DC, 22 May 2013

Council of European Energy Regulators (2013) Status review of renewable energy and energy efficiency support schemes. C12-SDE-33-03

Couture TD, Cory K, Kreycik C, Williams E (2010) A policymaker's guide to feed-in tariff policy design. National Laboratory of the U.S. Department of Energy, Technical Report NREL/TP-6A2-44849

Delbono F, Denicolo V (1991) Incentives to innovate in a cournot oligopoly. Q J Econ 106(3):951–961 (Published by: Oxford University, No. 3 (Aug., 1991)

European Commission (1997) Communication (1997) 599 final on energy for the future: renewable sources of energy. White Paper for a Community strategy and action plan

European Commission (2013) Communication staff working document SWD (2013) 440 final on guidance on the use of renewable energy cooperation mechanism

European Commission (2001) Directive 2001/77/EC of the European Parliament and of the Council of 27 September 2001 on the promotion of electricity produced from renewable energy sources in the internal electricity market. Off J Eur Union L 283:33–40

European Commission (2003) Directive 2003/96/EC of 27 October 2003 restructuring the community framework for the taxation of energy products and electricity. Off J Eur Union L 283:51–70

European Commission (2014) IP/14/400 State aid: Commission adopts new rules on public support for environmental protection and energy

European Renewable Energy Council (EREC) (2014) Costs of renewable energy sources (RES). EREC Factsheet

Grübler A, Nakićenović N, Victor DG (1999) Dynamics of energy technologies and global change. Energy Policy 27(5):247–280

Hayward JA, Graham PW, Campbell PK (2011) Projections of the future costs of electricity generation technologies. CSIRO Energy Transformed Flagship, Newcastle

Held A, Haas R, Ragwitz M (2006) On the success of policy strategies for the promotion of electricity from renewable energy sources in the EU. Energy Environ 17(6):849–868

Lee T, Wilde L (1980) Market structure and innovation: a reformulation. Q J Econ 94(2):429–436

McDonald A, Schrattenholzer L (2001) Learning rates for energy technologies. Energy Policy 29(4)255–261

Montgomery F (1943) Increased productivity in the construction of liberty vessels. Mon Labor Rev 57(5):861–864

Reichenbach J, Requate T (2012) Subsidies for renewable energies in the presence of learning effects and market power. Resour Energy Econ 34(2):236–254

Wright TP (1936) Factors affecting the cost of airplanes. J Aeronaut Sci 3:122–138

Yelle LE (1979) The learning curve: historical review and comprehensive survey. Decis Sci 10(2):302–328

Chapter 4
Strategic Choices for Sustainability at EU Regional Level

Today, the supply of electricity still comes mainly from large power plants fueled primarily by fossil fuels that operate through established distribution and transmission systems. Although these systems provide an efficient service worldwide for many years, times are changing. Modern societies have understood that in order to mitigate climate change, it is needed to reduce polluting emissions. The optimum use of traditional sources must be accompanied by an ever-increasing use of renewable energy sources (RES). There is therefore the spread of a wide range of energy sources that involve a number of complexities in terms of improvements in RES' technologies and design of electrical networks. The EU has launched several initiatives to increase the use of RES in all EU countries. With the climate and energy package, energy policy becomes a unilateral commitment in Europe. The achievement of macro-objectives outlined in the package is entrusted to the synergistic effects of different proposals closely linked. With regard to the RES Directive 2009/28/EC on the promotion of the use of energy from RES, it sets the overall policy framework in order to ensure coverage of 20 % of the energy demand of the EU through RES to the electricity, transport (biofuels), and heating–cooling sectors (European Commission 2009a). The objectives of mandatory coverage with RES are implemented at the national level and they vary from one country to another; they are between 10 and 49 % of final consumption by 2020.

Therefore, in this chapter we offer an updated and comprehensive review in which each EU country stands in relation to 2020 RES' objectives. The EU has a huge potential to develop a new generation of energy technologies like offshore wind, solar energy, or technologies for the exploitation of biomass, and research and development activities go ahead to get to possible RES technological breakthroughs. The EU is currently making a number of changes to the organization of the energy market, including rethinking the management of electrical energy. The influence of climatic conditions on the availability of solar and wind energy, together with the need to develop distributed systems, imposes the need of local networks capable of receiving and delivering electrical energy. Not having been designed to meet the needs of RES, grids need to be revised; a global and

S. Bigerna et al., *The Sustainability of Renewable Energy in Europe*,
DOI 10.1007/978-3-319-12343-1_4

widespread evolution is necessary. To achieve the objectives of energy security and emission reduction, smart grid technologies have attracted a lot of interest and the development of smart grids is becoming a priority of energy policy for the EU (Giordano et al. 2012).

The chapter proceeds as follows: Section 4.1 provides a description of RES allocation plans for EU countries. Section 4.2 describes where EU stands in the diffusion of RES technologies. Section 4.3 provides an overview of research activities in place to improve RES' productivity and Sect. 4.4 sets out to define smart grids and their penetration in Europe. Section 4.5 concludes the chapter.

4.1 Renewable Energy Action Plans in EU Countries

The Directive 2009/28/EC on the promotion of the use of energy from RES, which is also part of the climate and energy package, sets the overall policy framework in order to ensure coverage of 20 % of the energy demand of the EU through RES. The Directive 2009/28/EC also referred to as "indicative trajectory" that EU member states must follow to ensure that national targets are achieved (European Commission 2009a). The key instrument is the adoption of a National Renewable Energy Action Plan (NREAP) that is a planning document that provides detailed guidance on the actions to be implemented to achieve, from June 2010 until 2020, the binding objectives for all EU countries. The target is not equally divided between the EU member states, but each has an individual target for the overall share of RES in energy consumption. The objectives are to be achieved through the use of energy from RES in the electricity, heating–cooling, and transport sectors. It follows an analysis of the NREAP of all EU countries.

Austria The Austrian NREAP divides the target of 34 % RES into 70.6 % renewable electricity, 32.6 % in heating and cooling, and 11.4 % RES in transport. The projections of RES electricity mainly involve a relatively large use of hydro, followed by wind onshore, biomass, and pumping. As for the heating and cooling sector, the goal of 32.6 % in 2020 is achieved by 28.1 % from biomass, 2.1 % from solar thermal, and 0.3 % from geothermal. The objective of 11.4 % RES in the transport sector is achieved through employing 80 ktoe of bioethanol, biodiesel 410 ktoe, and 272 ktoe of RES electricity and 94 from other sources.

Belgium The Belgian NREAP divides the target of 13 % RES in 20.9 % renewable electricity, 11.9 % in heating and cooling, and 10.1 % RES in transport. The NREAP plan to achieve the mandatory target of 13 % RES by RES technologies that will produce approximately 5374.4 ktoe in 2020. The Belgian RES Federation, European development opportunities in rural area (EDORA), estimates that a target of 5985 ktoe (10 % more than the target NREAP) is economically achievable and is feasible. As for the heating and cooling and transports sector, the objectives of the NREAP and EDORA are quite similar, the predictions are much more ambitious for the electricity sector, according to which EDORA expects 26.6 % of renewable electricity in electricity consumption in 2020. The projections for RES % in electricity consumption include the use of wind onshore, biomass, and biogas. As for the heating and cooling sector, the NREAP projections for 2020 expect that the 9.1 % comes from biomass, 0.3 % from biogas, heat pumps from 1.6 %, and 0.9 % from solar thermal. The transport sector in 2020 rests essentially on biodiesel (698 ktoe), followed by renewable electricity (97 ktoe) and ethanol (91 ktoe). In order to achieve and exceed the target of 13 % RES in 2020, EDORA believes that it is crucial to strengthen the electricity grid to allow the development of power plants that produce energy from RES.

Bulgaria The Bulgarian NREAP divides the 16 % goal of energy from RES in 21 % renewable electricity, 24 % in heating and cooling, and 7.8 % RES in transport. The problems that Bulgaria has in relation to the grid connection procedures are: there are too many authorities involved (five different authorities); investors are not informed about the information flows between various authorities; there is a connection fee per Megawatt (MW), which depends on where the installation is made; network owners often deny network access to projects that have been licensed. There are some problems facing the RES, but they are not taken into account by the NREAP. For example, the permit procedures do not have a satisfactory legal framework. Many acts and ordinances have not been created specifically for the RES. In addition, the permit procedures differ from one municipality to another, creating contradictory rules. Existing networks should therefore be extended and modernized smart grids should be introduced to allow greater penetration of RES. The objective of 20.6 % from RES in electricity consumption should be achieved by using mainly hydropower, then wind onshore, biomass, and finally a small amount of photovoltaic. As for the heating and cooling sector, 23.8 % in heat consumption is achieved primarily through biomass (22.75), then biogas (0.4 %), solar thermal (0.5 %), and geothermal (0.2 %). The transport sector has difficulties to reach the target of 7.8 % of RES; however, it is expected that the target will be achieved with a mix of biofuels (196 ktoe), renewable electricity (5 ktoe), and others (4 ktoe).

Cyprus The Cypriot NREAP plans to achieve the overall target of 13 % RES by 16 % renewable electricity, 23.5 % in heating and cooling, and 5 % RES in transport. With regard to the electricity sector, the government of Cyprus considers that the law implemented by the RES Directive can restore investment security for most of the RES industry and especially for photovoltaic technology. The RES scenario in electricity sector should be obtained by using wind onshore, photovoltaic, and solar thermal electricity. As for the heating and cooling sector, it is more difficult for the target to be reached because there are a very limited number of applicants interested in installing RES for heating due to the lack of incentives encouraging appropriate publicity of support schemes. It is expected to reach the target of 23.5 % in heat consumption using 5.7 % biomass, 0.6 % of RES from heat pumping, and 17.2 % solar thermal. With regard to the share of RES in the transport sector, the objective of the government is very conservative since it only aims to 5 % and it should be achieved using only net biofuels imports (71.6 ktoe).

The Czech Republic The Czech NREAP divides the objective of RES to 13 % in 14 % renewable electricity, 14 % in heating and cooling, and 10.4 % RES in transport. The objective of 14 % RES in electricity consumption is split into biomass, biogas, hydro, photovoltaic, and wind onshore. One obstacle to the development of RES in the Czech Republic is the tendency to go into the hands of the state management of the national transmission network and the networks of regional distribution. The NREAP Czech is not very ambitious regarding the development of the smart grid even just because it does not develop the issue of smart grid. The heating and cooling sector does not expect the development of a scheme that improves the efficiency of the RES in the sector to 2020. The objective of 14.3 % RES in heat consumption should be achieved by implementing biomass (12.6 %), biogas (0.9 %), solar thermal (0.1 %), and geothermal (0.1 %). The action plan for transport in 2020 specifies that the 10.4 % RES in transport might be achieved mainly with biodiesel (495 ktoe) and bioethanol (128 ktoe).

Denmark The Danish NREAP has divided the 30 % overall target of energy from RES in 51.9 % in electricity consumption, 39.6 % in heat consumption, and 10 % in transport consumption. The Danish government considers that the 2020 target is feasible. This projection is based on the assumption of a political framework that gives incentives for RES. With regard to the electricity sector, in the RES the most widely used is wind power, both onshore and offshore; financial incentives for private investors are such that public authorities expect the wind power to grow at a fast pace. Biomass as wood chips and wood pellets imported should make an important contribution as fuel in heat and power plants. The 39.6 % in heat consumption should be achieved by using 32.4 % from biomass, 4.8 % from heat pumps, 2.2 % from biogas, and 0.2 % from solar thermal. As for the heating and cooling sector, the government has planned new initiatives to reduce energy consumption in existing buildings and in the new ones. In particular, energy consumption in new buildings must be reduced by at least 75 % by 2020 compared to the standards of 2009. Moreover, if it is feasible from an economic point of view, every new roof in buildings that consume more than 200 L of hot water per day must provide for the installation of solar heating. The transport sector will be covered by biodiesel (167 ktoe) and ethanol (94 ktoe); consumption of biofuels will be supported by imports from abroad.

Estonia The Estonian NREAP divides the target of 25 % RES in 17.6 % RES electricity, 38.4 % RES heating and cooling, and 9.9 % RES used in transport. The Estonian government believes it can achieve the goal to 2020 easily, especially since Estonia has a good potential in the field of RES. The 17.6 % in electricity consumption should be reached by employing mainly wind onshore and offshore and biomass. The promotion of RES in the heating and cooling sector is limited to small cogeneration installations and the reconstruction and modernization of district heating systems and it is reached by using only biomass. Transport sector should be covered by biodiesel (51 ktoe) and bioethanol (38 ktoe).

Finland The Finnish NREAP expected to reach the RES target of 38 % by the strategy of 32.9 % in electricity consumption, 47.5 % in heat consumption, and 20 % in transport consumption. Both the overall and sectoral targets are challenging, given the current system of support for RES; this means that Finland needs to revise a new system of incentives. The RES scenario for electricity involves the use of hydro, biomass, and wind. The heating and cooling sector projects to use 25.7 % biogas and 17.5 % biomass to reach the overall 47.5 % in heat consumption. RES used in transport will be mainly biodiesel (430 ktoe) and bioethanol (130 ktoe); furthermore, it is provided for the use of electric vehicles (40 ktoe).

France The French NREAP splits the 23 % RES target in 27 % in electricity consumption, 33 % in heat consumption, and 10.5 % for RES used in transport. The electricity sector is driven mainly by large hydro and wind power; photovoltaic technology has a rather slow development. There have been some changes in the regulation concerning RES that have affected investment in the sector. The heating and cooling sector is estimated that the measures taken in the construction industry are stimulating for the installation of RES technologies. The 33 % goal in heat consumption is reached making use of biomass (26.5 %), geothermal (4 %), solar thermal (1.6 %), and biogas (0.9 %). In the transport sector, all EU member states are obliged to draw up a national scheme by which national operators demonstrate the sustainability of their biofuels or bioliquids. At the time of writing NREAP, France did not show any scheme. The projections to 2020 for RES used in transport expect to implement biodiesel (2850 ktoe), bioethanol (650), renewable electricity (402 ktoe), and others (110 ktoe).

Germany The German NREAP intends to achieve the target of 18 % RES by implementing the portfolio of 38.6 % RES in electricity consumption, 15.5 % RES in heat consumption, and 13.2 % RES in transport. Although the German government expects to achieve the goal of 19.6 % of RES in gross final energy consumption in 2020, the RES target remains at 18 %. The electricity sector is driven by wind and solar power; the sector is affected by the cuts made to the system to support RES and there is concern among investors with respect to the long-term stability of the support policy. The NREAP does not have very ambitious plans for existing and new buildings; the overall objective of 15 % in heat consumption should be reached by 10.4 % biomass, 1.8 % biogas, 1.3 % solar thermal, 1.2 % RES from heat pumps, and 0.8 % geothermal. The transport sector is the one that, according to the NREAP, has the greatest need to be revitalized; the increase in consumption of biofuels will be satisfied by imports, with negative impacts on both the domestic industry and the environment. The projections for RES in transport expect to employ biodiesel (4443 ktoe), bioethanol (857 ktoe), RES electricity (667), and others (173 ktoe).

Greece The Greek NREAP has divided the 18 % overall target of energy from RES in 39.8 % in electricity consumption, 19.7 % in heat consumption, and 10.1 % in transport consumption. The RES electricity scenario intends to achieve the goal by using mainly wind power even if the NREAP notes that the share of RES technologies should be reviewed in order to have a better balance between various RES. The heating and cooling sector limps unless additional measures will be taken; the overall objective should be reached by making use of biomass (12.6 %), solar thermal (3.7 %), and geothermal (0.5 %). In the transport sector, the Greek government is more active and regulatory action plans to promote domestic production of biodiesel; a lot of effort will be put on exploring the local potential and the development of delivery channels. The projections for RES in the transport sector involve the use of bioethanol (414 ktoe), biodiesel (203), and RES electricity (16.5 %).

Hungary The Hungarian NREAP splits the target of 13 % RES in 10.9 % RES electricity, 19.2 % RES heating and cooling, and 11.2 % RES used in transport. The Hungarian government is expected to exceed the overall target of about 1.7 % by 2020. The electricity target should be reached by making use mainly of biomass and wind. RES' projections for heating and cooling take into account biomass (12.6 %), geothermal (3.7 %), solar thermal (0.8 %), and biogas (0.6 %). Projections for RES used in transport make use of biodiesel (295 ktoe), bioethanol (189 ktoe), RES electricity (24 ktoe), and biofuels (22 ktoe).

Ireland The Irish government has divided the 16 % RES overall target of energy in 42.5 % from electricity consumption, 12 % in heat consumption, and 10 % in transport consumption. The Irish government has in its NREAP also an export scenario, given the great potential of Ireland in terms of RES. The electricity sector must be supported by effective policies and a stable market has to be created for investors so that Ireland really manages to become even an exporter of energy from RES. Projections for RES electricity in 2020 are supported mainly by wind power and biomass. The outlook for the heating and cooling Irish sector is smaller; the domestic policy gives a subsidy to homeowners in terms of contribution to the cost of installing an RES technology in the existing home. The objective in heating and cooling should be reached by using biomass (9.2 %), RES from heat pumping (1.7 %), biogas (0.7 %), and solar thermal (0.4 %). With regard to the transport sector, the national scheme of 2010 requires all suppliers of road fuel to use biofuels in the fuel mix initially for an amount equal to 4 % per year, to be increased over time. It is expected to achieve the target with biodiesel (342 ktoe) and bioethanol (139 ktoe).

Italy The Italian NREAP plans to achieve the overall RES target of 17 % through the use of 26.4 % in electricity consumption, 17.1 % in heat consumption, and 10.1 % in transport consumption. Regarding electricity, Italy uses a fairly broad mix of RES from hydro and wind energy to photovoltaic, etc. However, the network is not yet capable of supporting the requirements for the connection of RES, especially in remote areas where there is a great potential for RES. In the field of heating and cooling, there are no specific incentives for the development of the sector, with the exception of some tax deductions. The 17.1 % in heat consumption is planned to be achieved through biomass (8.8 %), RES from heat pumps (4.8 %), solar thermal (2.6 %), geothermal (0.5 %), and biogas (0.4 %). The diffusion of RES in transport in Italy is based on a mandatory minimum share of biofuels which is

3.5 % of the total fossil fuel put into the market in 2010. RES projections for RES used in transport involve the use of biodiesel (1880 ktoe), bioethanol (600 ktoe), and RES electricity (369 ktoe).

Latvia The Latvian NREAP expects to reach the RES target of 40 % by 59.8 % in electricity consumption, 53.4 % in heat consumption, and 10 % in transport consumption. The electricity sector has as a main barrier to the development of RES the access to the network, which is not priority for RES and the expense to access the network lies with the RES developers. RES electricity sector lies mainly on hydro and biomass. The heating and cooling sector is supported by new tax measures and projects are funded by European structural funds in the form of direct funds that contribute to approximately 25 % of total costs. The 53.5 % goal in heat consumption consists of 51.4 % biomass, 1.9 % biogas, and 0.1 % solar thermal. In the transport sector, Latvia promotes biofuels through direct support measures and fiscal measures. Most biofuels are derived from rapeseed oil. RES used in transport consumption are biodiesel (28 ktoe), bioethanol (18 ktoe), RES electricity (6 ktoe), and others (31 ktoe).

Lithuania The Lithuanian NREAP has divided the 24 % overall target of energy from RES in 21.3 % in electricity consumption, 39.2 % in heat consumption, and 10 % in transport consumption. The electricity sector is driven primarily by wind, although the Lithuanian government plans to boost solar and cogeneration installations with appropriate policies. The Lithuanian NREAP with regard to the heating and cooling sector does not provide any specific measures of support, but only one set of laws and rules that must be implemented in the coming years. Biomass (36.3 %), biogas (1.9 %), RES from heat pumps (0.5 %), solar thermal (0.3 %), and geothermal (0.2 %) will contribute to achieving the expected quota of 39.2 % in heat consumption. The transport sector will be driven by biodiesel (131 ktoe), bioethanol (36 ktoe), and RES electricity (3 %).

Luxembourg The Luxembourgish NREAP plans to achieve the overall target of 11 % RES by 11.8 % renewable electricity, 8.5 % in heating and cooling, and 10 % RES in transport. The RES electrical industry should be driven mainly by wind and biomass. With regard to the heating and cooling sector, the NREAP provides a list of possible tools to support the sector, but there are no specific policy measures that aim at ambitious growth rates. The 8.5 % in heat consumption is reached through the use of biomass (5.5 %), RES from heat pumps (1.3 %), biogas (1.1 %), and solar thermal (0.6 %). According to the NREAP, the percentage of RES in the transport sector will increase by 2.1 % in 2010 to 10 % in 2020, through the use of bioethanol (23 ktoe) and biodiesel (193 ktoe).

Malta The Maltese NREAP splits the target of 10 % energy from RES in 13.8 % RES electricity, 6.2 % in heating and cooling, and 10.7 % RES in transport. The electricity industry scenario with regard to RES with priority is based on wind and biomass. The Maltese government has not introduced specific measures in relation to the RES and the construction of new buildings, leaving most of the initiative at the discretion of individuals. The projections in heat consumption refer to solar thermal

(3.8 %) and biogas (2.4 %). The most important challenge of the NREAP Maltese regards the transport sector, so that the contribution of biofuels makes strides year after year. RES in transport sector to 2020 rests on biodiesel (7.03 ktoe), bioethanol (5.79 ktoe), and RES electricity (0.7 ktoe).

The Netherlands The Dutch NREAP expects to reach the target of 14 % of energy from RES by 37 % in electricity consumption, 8.7 % in heating and cooling sector, and 10.3 % in transport consumption. The electricity sector needs greater stability and certainty in the long run to reach the target set for 2020 and go further. The projections for RES electricity in 202 are driven by wind and biomass. The heating and cooling sector was not a priority in Dutch NREAP but it is expected to receive more attention from the Dutch public authorities. Total RES percentage in heat consumption will be reached by implementing biogas (3.5 %), biomass (2.6 %), RES electricity (1.5 %), geothermal (1 %), and solar thermal (0.1 %). Policy in the transport sector is the most rigorous: since 2007, it is mandatory to mix the fuel for road with biofuels; in 2010, the percentage was 5.75 %. RES projections in transport consumption concern biodiesel (552 ktoe), bioethanol (282 ktoe), and RES electricity (71 ktoe).

Poland The Polish NREAP expects to reach the RES target of 15 % by 19.1 % in electricity consumption, 17.1 % in heat consumption, and 11.7 % in transport consumption. The RES industry scenario in electricity sector mainly consists of wind and biomass. With regard to the heating and cooling sector, the RES portfolio to achieve the objective is composed of biomass (13.4 %), solar thermal (1.5 %), biogas (1.3 %), geothermal (0.5 %), and RES electricity (0.4 %). The biofuels in the transport sector are driven by policies of exemptions and fee reductions; RES projections in transport sector rest on biodiesel (1451 ktoe), bioethanol (451 ktoe), RES electricity (50 ktoe), and others (66 ktoe).

Portugal The Portuguese NREAP splits the target of 31 % energy from RES in 55.2 % RES electricity, 30.6 % RES heating and cooling, and 10 % in the transport sector. Portugal has a great potential in terms of RES, but the industry is not fully exploited and therefore there is need to review the criteria and administrative procedures of the country in relation to the RES. As described in the Portuguese NREAP, the goal in the electricity sector should be achieved by employing for the most part wind and hydro. The heating and cooling sector is affected by a revision of the rules for construction of buildings in favor of RES; surprisingly, Portugal expects a decrease in the contribution of biomass due to the introduction of more efficient burners. The 30.6 % RES goal in heat consumption consists of biomass (27.9 %), solar thermal (2 %), biogas (0.4 %), and geothermal (0.3 %).

Romania The Romanian NREAP has shared the 24 % overall target of energy from RES among 42.1 % in electricity consumption, 22 % in heat consumption, and 10 % in transport consumption. The electricity sector's projections are driven mainly by hydro and wind power. RES in the transport sector are promoted by the Romanian government according to which, since 2010, all types of gasoline and diesel fuel sold in the market must have a minimum percentage of biofuels that amounts to 5.75 %;

projections to 2020 refer to biofuels (489 ktoe) and RES electricity (14 ktoe). With reference to the heating and cooling sector, Romania had not yet a framework for heating and cooling sector.

Slovakia The Slovak NREAP expects to reach the RES target of 14 % by 24 % in electricity consumption, 14.6 % in heat consumption, and 10 % in transport consumption. The electricity sector is characterized by a continuous change in legislation that adversely affects the sector; it is crucial to increase the confidence of investors. Projections for RES electricity in 2020 take into account the development of hydro, biomass, and wind power. The heating and cooling sector lacks concrete measures to promote the use of energy from RES. Total RES percentage in heat consumption should be reached by implementing biomass (11.2 %), geothermal (1.6 %), biogas (1.1 %), solar thermal (0.5 %) and RES from heat pumps (0.2 %). Since 2007, Slovakia has introduced a scheme for the reduction of excise tax on biofuels; the 10 % RES in transport consumption will be achieved by biodiesel (110 ktoe), bioethanol (75 ktoe), and RES electricity (17 ktoe).

Slovenia The Slovenian NREAP divides the 25 % goal of energy from RES in 39.3 % renewable electricity, 30.8 % in heating and cooling, and 10.5 % RES in transport. The electricity sector is driven in large part from hydro and the other RES are present for small units. The target for 2020 relative to the heating and cooling sector should be easily accessible because Slovenia is rich in wood and other biomass; projections for RES heating and cooling in 2020 consist of biomass (25.9 %), RES from heat pumps (2.9 %), solar thermal (1 %), and geothermal (1 %). The situation is different and more complicated for the transport sector because it starts in 2008 by the presence of only 1.22 % of biofuels; there is a need to introduce policies to reduce traffic on the road and the rail system needs to be modernized. By the way, projections for RES in transport go by biodiesel (173.7 %), bioethanol (18.5 %), and RES electricity (10.5 %).

Spain The Spanish NREAP splits the 20 % RES goal in 40.2 % renewable electricity, 18.9 % in heating and cooling, and 13.6 % RES in transport. The electricity sector has been affected by a series of cuts that have created difficulties for the development of RES in Spain and have given the image of a lack of stability in the long term. The mix of RES in electricity sector is quite varied, although wind and hydro are the masters. With regard to the heating and cooling sector, the 18.9 % goal should be reached through the use of biomass (16.2 %), solar thermal (2.2 %), biogas (0.3 %), and RES from heat pumps (0.2 %). The transport sector's projections to 2020 provide the use of biodiesel (2900 ktoe), bioethanol (400 ktoe), RES electricity (380.9 ktoe), and others (4 ktoe).

Sweden The Swedish NREAP expects to reach the RES target of 49 % by 62.9 % in electricity consumption, 62.2 % in heat consumption, and 13.8 % in transport consumption. Sweden has already a high share of RES in its energy portfolio in 2009, so the overall target is very high. The projections for electricity sector with reference to RES are based on hydro, biomass, and wind. The conditions of RES in heating and cooling sectors are good; bioenergy is the largest source of energy used and there are plans to use biomass to a greater extent. The 62.2 % objective in heat consumptions is based on the use of biomass (55.9 %), RES from heat pumps (6.2 %), and biogas (0.1 %). Policy measures to increase RES in the transport sector are smaller; the Swedish government has decided to add that only 5 % of ethanol to gasoline, in contrast to many other countries who came to 10 %.

The United Kingdom The UK NREAP expects to reach the RES target of 15 % by 31 % in electricity consumption, 12 % in heat consumption, and 10.3 % in transport consumption. The RES electricity scenario to 2020 is based mainly on wind. Projections for RES heating and cooling refer to biomass (7 %), RES from heat pumps (4.4 %), biogas (0.6 %), and solar thermal (0.1 %). The 10.3 % RES in transport consumption will be achieved by biodiesel (2462 ktoe), bioethanol (1743 ktoe), and RES electricity (267 ktoe).

4.2 RES Targets: Where Are We Going?

Each technology has its own characteristics with advantages and disadvantages. To ensure a reliable supply of electricity, there should be a balanced mix of different technologies. First, there are needed plants that ensure energy band, but it is necessary to have also flexible systems, capable of storing excess energy produced by wind turbines and photovoltaic panels, to use it during peak consumption. To generate energy, EU countries are resorting to different sources, depending on the primary energy reserves available in the area or on the basis of certain political principles.

4.2.1 Solar Thermal in EU Countries

The solar thermal sector is consolidating in the EU with consistent growth rates and encouraging results. Although the development of the sector in recent years in the EU has scored good results, there is a little slowdown due to the current financial crisis that also involves the solar thermal sector whose investment costs are still quite high. Therefore, the surface of solar thermal panels installed in the EU in 2012 amounts to 3,395,420 m^2, a slight decrease of 5.5 % compared to 2011 with 3,591,580 m^2. The situation is positive in some EU countries: the French market was able to fight through the development in the field of condominiums. The solar thermal market is growing in Greece, Poland, Hungary, and Denmark, as a result of rising energy prices (gas and heating oil). In particular, Poland remains one of the countries in which the solar heat gains market shares: the country has become the third largest market in the EU European exceeding the threshold of 300,000 m^2. The reasons for the growth are the strong increase in the price of gas from Russia and the success of the incentive program funded by the National Fund for Environmental Protection and Water Management.

The success of the Danish solar thermal market is particularly envied at the EU level: in 2012, the market has risen to 133,122 m^2, i.e., more than double (113 %) compared to 2011 (62,401 m^2). The Danish solar thermal market is unique in that it is essentially based on the development of solar installations for condominiums and fields of solar thermal collectors to supply district heating networks (122,000 m^2 in 2012).

Markets in Belgium, the Netherlands, and Luxembourg are growing, but not enough to reverse the trend in the EU. The main reasons for the poor state of health of the European market solar thermal are the same for a few years (the recession and stagnation in the construction sector), but other technologies for the production of renewable energy, such as photovoltaic and heat pumps, have competed in the sector.

Germany, the only EU country to install over 1 million m^2 per year, is no longer the leading country. In southern Europe (Spain, Italy, and Portugal), where there is the greatest potential for solar thermal, the continuing economic crisis combined with the collapse of the construction sector is stifling the development of the solar thermal sector in spite of the implementation of favorable technical standards.

The Austrian market, the showcase of the technology of solar thermal in the EU, is declining: the country has installed only 208,800 m^2 in 2012, down from 235,889 m^2 in 2011. This decrease is a source of concern, given that the majority of provinces have increased their incentives in 2012 to revive the market.

So, the German and Austrian markets have not gone well in 2012 also perhaps because of the weather variable that has not encouraged the development of this technology. A reversal of this trend is also difficult because the solar thermal is already well developed in these two countries and the prospects for growth logically diminish over the years and with the increase in the number of plants. The potential for growth is much higher in the markets of France and Italy, but the financial and economic crisis that is currently hitting these countries represents a major obstacle to the development of the market.

As the Polish market, some markets are better resistant thanks to the sharp rise in energy prices, but this is only part of the explanation. The markets reflect the trends of the construction market and the financial capacity of households.

By considering the current pace of growth of the production of solar thermal energy, that is 7.7 % between 2011 and 2012, by assuming to maintain this rate until 2020, the EU will achieve only half of the cumulative results of NREAP relative to solar thermal. According to the cumulative NREAP of all EU countries, the production of solar thermal would amount to 3019 ktoe in 2015 and 6348 ktoe in 2020; the current trend instead estimates the production of solar thermal to be 2156 and 3125 ktoe in 2015 and 2020, respectively.

A change of scale is only possible if there will be a greater political will of the authorities. In this sense, it should also be noted that Directive 2010/31/EC on energy performance of buildings provides for new buildings energy consumption close to zero by 2020 (and by 2018 for public buildings) and covered by the energy produced from RES. It also sets minimum energy performance requirements for buildings under construction or for existing ones undergoing a major renovation.

The incorporation of these directives, with high-enough constraints on older buildings, will allow the industry to face new challenges both in terms of production cost and in terms of efficiency of the systems.

4.2.2 Photovoltaic in EU Countries

From 2010 to 2012, the growth of photovoltaic in the EU has been mainly driven by speculative investments that took advantage of the difference between the levels of guaranteed remuneration and the very rapid decline in the production's cost. Some governments are trying to affect costs by introducing taxes on the production of electricity or retrospectively by amending their laws. The Czech Republic has been the first country to take the first steps in this direction: by the end of 2010, it has adopted a tax on investments made previously between 2009 and 2010. Bulgaria has introduced a tax on access to the grid for the systems entered service from April 2010. Greece has introduced at the end of 2012, a tax of up to 30 %, on the proceeds of the systems already installed or systems that will be installed in the future. Belgium has also moved in this direction and Spain has introduced a 7 % tax on the earnings of all electricity producers. The European Photovoltaic Industry Association has asked the EU to take measures against countries that fail to fulfill their commitments in terms of support for RES. Some markets have had a rather positive development of the PV market: France and Greece have passed the threshold of 1 GWp. In particular, France has connected to the network 1079 MWp in 2012 and the overall photovoltaic park connected to the grid has risen to 4003 MWp. In order to stimulate the market, in January 2013, the French government has put in place new rules through two measures related to a change in the tariff system and a bonus on the tariffs for European manufacture modules. The UK market is not particularly dynamic in 1655 MWp in 2012 connected to the network; for this reason, the British regulator for the electricity and gas markets (Ofgem) has decided to maintain stable support measures to photovoltaics. The Danish and the Dutch market took off in 2012, thanks to the success of the net metering.[1] Denmark is, in fact, one of the few EU countries to have had a significant growth of its fleet in early January 2013, having exceeded the threshold of 400 MWp, thanks to the system of net metering by allowing individuals with photovoltaic systems to offset their electricity consumption with their productions on an annual basis. All consumption in compensation during the year is exempted by the payment of the electricity fee and network fee and also by very high taxes on electricity consumption in Denmark. The installed capacity is also increased in the Netherlands with additional 175 MWp in 2012, almost three times that of 2011, with a total capacity connected to the grid in 2012 that equals 316 MWp. Germany has surpassed its own record of installations, having connected to the network 7604 MWp in 2012 compared to

---

[1] Net metering is a mechanism that allows to enter into the grid the net electricity generated by a system of private production of electricity, but not immediately self-consumed, and then pick the energy up at a later date to satisfy its electricity consumption (Edison Electric Institute 2013).

7485 MWp in 2011. Photovoltaic energy currently accounts for 4.6 % of the gross production of electricity in Germany, after the wind (7.3 %) and all sectors of biomass (5.8 %), but ahead of hydropower (3.3 %). In 2012, the photovoltaic market in Italy has exceeded expectations with an aggregate photovoltaic capacity connected to the grid of 16,350 MWp at the end of 2012, i.e., 3577 MWp more than 2011. The cumulative power in the EU in 2012 was 68,647 MWp: the production is currently three times higher than that of 2010 and photovoltaic energy accounts for over 2 % of electricity consumption in the EU. The forecast for 2020 is quite optimistic. The European Photovoltaic Industry Association, in its publication in September 2012, "Connecting the sun" shows a contribution of 12 % of photovoltaic in 2020 and 25 % in 2030 to EU demand. In comparison, the scenario of NREAP includes a contribution of 2.4 % in 2020, equivalent to 83.4 TWh of production and an installed capacity of 84,376 MW. Actually, with current trends, it should be reached the threshold of 100 GWP by 2015.

4.2.3 Wind Power in EU Countries

The wind power installed and connected to the grid in the EU in 2012 has been 11,840 MW, enabling the EU to exceed the threshold of 100 GW of installed capacity (105,635 MW). In 2012, the major producing countries of wind were Spain, Germany, UK, Italy, France, Portugal, and Denmark. In particular, the first three countries have provided 57.7 % of the EU wind power production. Germany has 2439.5 MW installed in 2012, bringing the wind power of the German park to 31,331.9 MW. The repowering of power plants allows to obtain an increase in power generation, amounting to 18% in the German market. In 2012, the UK has connected more wind power offshore and onshore. In 2012, the UK has increased its wind power in operation of 1853.9 MW, of which 1156.4 MW is offshore. Spain has seen a wind power capacity in 2012 that amounted to 22,579 MW, compared to 21,239 MW in 2011 and it is the second EU country for wind power installed. In 2012, Italy has been the fourth EU wind energy market; according to the National Association of Wind Energy (ANEV), Italy has installed 1272 MW in 2012, bringing the installed power of Italian wind farm to 8144 MW. In France, the situation is less thriving: the connection of new wind power in 2012 amounted to 757 MW, which corresponds to the third consecutive year of decline in the market. The reason for this decline is to be found in the accumulation of bureaucratic and administrative procedures that have slowed the growth of the sector. However, France has one of the largest wind potential in Europe and then the wind should be further exploited. Poland is currently one of the most active markets in the EU; according to the Institute for Renewable Energy, the country has installed 884 MW during 2012, bringing the power of Polish wind farm to 2500 MW. Overall in 2012, the EU wind energy market has exceeded expectations, although as early as 2013 there was a slower pace due to the economic uncertainty that has characterized the EU and led a number of governments to revise their incentive systems or to not provide sufficient guarantees to investors. Considering the NREAP, the EU is currently in the middle of the road on its target installations in 2020, corresponding to 213,563 MW including 44,244 MW offshore. For 2012, the European Wind Energy Association has calculated that the trajectory of the NREAP was 107,602 MW, with a slowdown in installations in 2013. In fact, at the end of 2013, the installed capacity of wind power was equal to 117.3 GW, of which 110.7 GW is onshore and 6.6 GW is offshore. The installed capacity of wind power in 2013 has been 11,159 MW, a decrease of 8% compared to the installations of 2012 (EWEA 2014). The situation is very different among EU countries: while some countries such as France, Spain, Portugal, and Greece are lagging behind in meeting their targets outlined in the NREAP, other countries such as Sweden, Italy, and Poland drag growth in the EU. Germany, the European leader, is currently in line with its goals, while the UK begins to stall. The EU is achieving good results with respect to its goal of installations in 2020 to 213,563 MW from wind power: installations in 2012 amounted to a level of 107,602 MW. The positive trend of wind power should be seen in the light of the intensive work of the EU for the development of a new framework that will help to modernize and expand EU's energy infrastructure.

4.2.4 Biofuels in EU Countries

In 2012, there has been a slight increase in the consumption of biofuels in the EU compared to 2011: consumption was around 14.4 Mtoe in 2012 compared to 14 Mtoe in 2011, i.e., an increase of 2.9%. The growth in 2012 was heterogeneous among EU countries. In particular, 14 countries have increased their consumption (e.g., France, Spain, Sweden, and Finland), 10 others (such as the UK, Poland, Hungary, and Italy) have decreased. Others do not have used biofuels in 2012 (Malta, Estonia, and Romania).

The drop in consumption occurred in some countries is due to two reasons: the economic crisis, which has led some importing countries to reduce its level of incorporation of biofuels, and the uncertainties about the new EU legislation. The breakdown of the consumption of biofuels has remained essentially the same as in previous years, with biodiesel representing 79.1% of total energy consumption in consumption, far ahead of bioethanol (19.9%). The pure vegetable oil and biogas account for 1% of total consumption. Germany is the country that consumes the largest amount of biofuels in the transport sector, with a total consumption in 2012 amounted to 3,018,321 toe, of which 73% is biodiesel and the remaining is bioethanol and to a lesser extent vegetable oil well. The totality of this consumption has been certified, allowing the country to account for them in order to reach the objectives in the field of RES. Germany then brings the official quota of biofuels on road to 5.7% in 2012 compared to 5.5% in 2011.

France is the first European consumer of biofuels, but in 2012 became the first consumer of biodiesel. The country has in fact consumed 2,299,800 toe of biodiesel in 2012 and 417,600 toe of bioethanol, bringing the total consumption to 2.7174 million toe. The consumption of biofuels in France is therefore increased by 12% from 1 year to the next. The rate of incorporation of transportation on the French mainland is 6.8%, one of the highest rates in Europe. In September 2012, the government filed a new action plan for agriculture by establishing that the incorporation rate of first-generation biofuels may not exceed 7%. The target of 10% should be achieved through business development, biofuels based on waste collection, waste, algae, or residual cellulose and through the development of electric vehicles or hydrogen powered.

The UK has experienced a substantial drop in the consumption of biofuels amounted to 15.9% between 2011 and 2012 (from 1056 ktoe in 2011 to 888 ktoe in 2012). Specifically, there have been used 634 million L of biodiesel (31% less than in 2011) and 775 million L of ethanol (19% more) in 2012. The consumption of biodiesel and bioethanol has lined up as the share of the energy content of biodiesel decreased from 72.3% in 2010 to 69.0% in 2011, reaching up to 56.2% in 2012. The Department of Energy and Climate Change attributes this English decline in the consumption of biodiesel in legislative changes from April 2012. Spain is among the countries that have maintained a strong growth in its consumption of biofuels in 2012, where in transport about 1719–209 ktoe of biodiesel and bioethanol was consumed. Spain thus retains the third position in the consumption of biofuels in the EU with a rate of incorporation in transport by 7.3% in 2012. Sweden has the highest rate of incorporation among all EU countries: the share of sustainable biofuels with respect to fuel consumption in the transport sector has increased from 6.3% in 2011 to 7.8% in 2012. Total consumption of biofuels in Sweden amounted to 586,887 toe in 2012, of which 207,564 is bioethanol and 307,929 is biodiesel. Sweden is definitely part of the country's most ambitious clean transport. In 2012, a program was initiated by the Swedish government in order to find solutions to ensure that Swedish vehicles are completely independent from fossil fuels by 2030.

The consensus forecast for total EU biofuels consumption in 2020 is in line with the NREAP of EU member countries. The cumulative amount of biofuels envisaged by NREAP in 2020 amounts to 29,745 ktoe (of which 7307 is bioethanol, 61,649 is biodiesel, and 789 others) and with the current trend it is expected to reach an amount equal to 30,000 ktoe.

4.2.5 Biomass in EU Countries

In 2011, primary energy production from solid biomass (wood, wood waste, other biomass plants or animals) has declined by 2.4 Mtoe, reaching 78.8 Mtoe. The gross consumption of primary energy from solid biomass, which includes imports and exports, has amounted to 80.8 Mtoe in 2011, a decrease of 3.9% over the previous year.

The main reason for this difference is the increase in imports of wood pellets from Canada and the USA. By looking at the statistics of the past two decades, the decline of energy production from solid biomass is particularly unusual given that the biomass in the EU has grown steadily since 1990, except for a slight decline in 1999. Indeed, it is more than doubled in the period 1990–2010 (39.5 Mtoe produced in 1990). As far as the production of heat in the processing, which corresponds to sales of district heating networks, there is a decrease of 7.5 % in 2011 for a production of 7 Mtoe. Much of this heat, 60.8 % in 2011, has been produced by cogeneration plants (plants that produce both heat and electricity). The final energy consumption is 58 Mtoe in 2011, representing a decrease of 3.1 % compared to 2010. The total heat consumption is 64.9 Mtoe in 2011, down from 2.4 Mtoe compared to 2010. The production of electricity from solid biomass has fared better in the EU since the electric heating is less common than other types; production amounts to 72.8 TWh in 2011, which represents an increase of 2.6 % compared to 2010. Sweden has had a decline particularly strong energy production from solid biomass: production from solid biomass is decreased by more than 1.7 Mtoe, reaching 8.2 Mtoe. In Sweden, biomass energy is used mainly by forest industry, which produces both heat for the residential sector, and electricity for their own consumption. In Finland, the decline in consumption has been less significant and the country has registered a primary energy production from solid biomass in 2011 that amounts to 7476 Mtoe. In Germany, the decline in the consumption of solid biomass can be attributed to the lower heating needs of families. Among the various biomass, wood pellets have been the one that have suffered less. According to the German Association of Energy and Forestry (DEPV), the pellet production has increased by 6 % in 2011 (plus 110,000 t), amounting to 1.86 million t. Germany stands as the first EU country to primary energy production from solid biomass, amounting to 11,690 Mtoe in 2011. The UK is one of the few EU countries to have increased the consumption of solid biomass (the country imports more than a quarter of its consumption). The only beneficiary of this increase, according to the Department of Energy and Climate Change, has been producing electricity, which has risen to 6.1 TWh, increasing by 16.9 % between 2010 and 2011. The British government has also implemented a system of self-certification for coal plants that are entirely converted into biomass power plants, or plants that produce co-combustion on a large scale for 2013–2017 period. The primary energy production from solid biomass in the UK amounts to 1756 Mtoe in 2011. As in Germany, in France the decrease of energy production from solid biomass (from 10.6 Mtoe in 2010 to 9.2 Mtoe in 2011) is due to the fall in household consumption. In Poland, the sharp increase in the consumption of solid biomass derives in large part by the rapid development of the co-combustion of biomass for the production of electricity. The electric power has consumed 1 million t of biomass in most in 2011 of which 90 % is used in 51 plants operating in co-firing. The volume of biomass measured in plants co-firing is increased by three times between 2006 and 2011 (from 1.7 to 5.2 million t), and a significant part of this biomass was imported. However, the Polish government has decided to reduce support for the biomass sector because this rapid development is due also to the strong economic

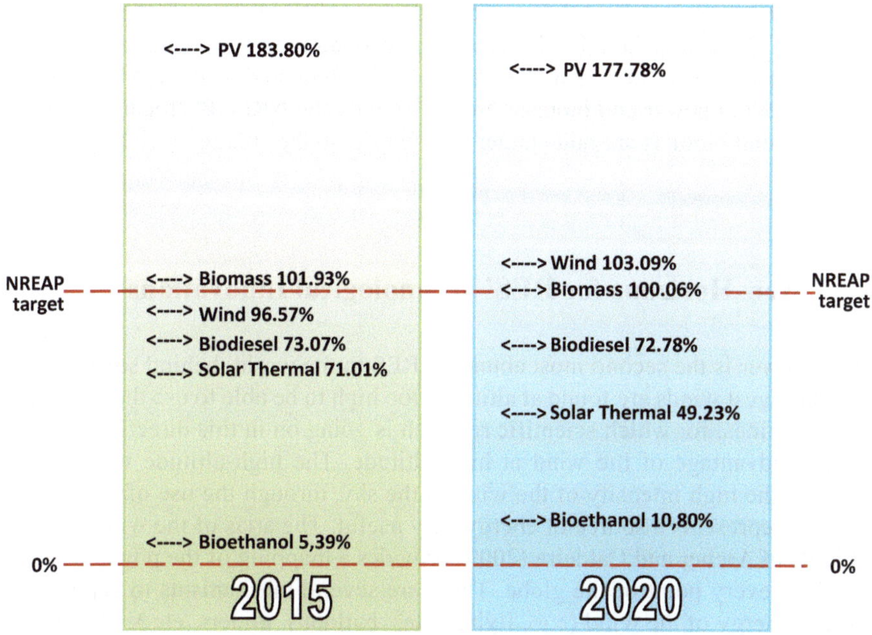

Fig. 4.1 Comparison between the current trend of RES and National Renewable Energy Action Plans (*NREAP*). (Source: our elaboration from EurObserv'ER 2013)

incentives that are likely to create an unsustainable use of biomass. The European industry of biomass was only slightly affected by the recession. The main driver of growth is the production of electricity, regardless of whether it happens or not in cogeneration. The majority of EU member states, driven from their commitments taken in the NREAP for RES, have created the context for the development of electricity production from biomass by applying incentive mechanisms for the production of electricity from RES and requiring operators to reduce their emissions of greenhouse gases (GHGs). This strong political will arrives at a favorable moment, when coal plants are aging and they require significant investment to make the modernization or construction of new power plants. Data on electricity production show that the EU is in line with the objectives of the NREAP; According to the NREAP roadmap, electricity production from biomass in 2020 should amount to 154.9 TWh and with the current trend it is estimated to get to produce 155 TWh.

In conclusion, the analysis carried out so far detects that some RES have sustained growth rates to the point that the installed amount will be greater than expected from the overall NREAPs (photovoltaic, wind, and biomass), while other RES can be found at just bottom of the rod outlined by NREAPs. Comparisons between the current trend of all RES analyzed in this chapter and NREAP of the EU both in 2015 and in 2020 are shown in Fig. 4.1.

Some RES, as described above, exceed the goal NREAP, first of all photovoltaic that, with the current trend, will install photovoltaic power for an amount greater than 183.80% in 2015 and 177.78% in 2020 compared to the roadmap of the NREAP. Wind power and biomass are in line with the NREAP's roadmaps, while solar thermal biofuels are quite under with respect to the cumulative NREAP's objectives.

4.3 New Horizons for RES' Technological Innovations

Wind power is the second most abundant RES in the world, behind solar energy. The strongest winds are found at altitudes too high to be able to use the stationary wind turbines, for which scientific research is going on in this direction in order to take advantage of the wind at high altitude. The high-altitude wind power exploits the high intensity of the wind in the sky, through the use of cable technology; represents a source of energy very useful. The atlas of the winds at high altitude of Archer and Caldeira (2009) provides a mapping of the power of these winds at every point of the globe. There are several mechanisms to capture the kinetic energy of the wind (e.g., flying kites, balloons, gliders, etc.). The energy generated by a wind power system of high altitude can be used in high or sent to the surface of the ground conducting cables of mechanical force through a tether.

The winds at higher altitudes are quite stable, persistent, and of greater speed, because the power available in the wind increases as the cube of the speed; they assume different parameters and twice the speed. With more stable and predictable wind, the wind at high altitude has an advantage over a wind near the ground. It is necessary to be able to identify the high-altitude wind using the vertical size of the airspace for the generation of energy.

In particular, the wind speed decays gradually when it descends toward the ground, as the air masses transform part of their kinetic energy into heat by friction between them and with the orography of the territory. However, for heights, not too high, wind regimes are still likely to greatly enhance the production of wind energy compared to units near the ground. For example, 80 m is the height representative of a wind generator to the ground, which therefore has a lower specific power at 30% of that at 800 m (Table 4.1).

The EU is determined to develop smart grids: in particular, the Parliament and the European Council have approved, as part of the so-called third energy package (European Commission 2009b, 2009c), obligations of installation of

Table 4.1 Effect of the height above the ground for wind power. (Source: Canale et al. 2009)

Height above the ground (m)	Wind speed (m/s)	Specific power (W/m^2)
800	7.2	205
80	4.6	58
10	3.3	22

smart electric meters and the request to EU member states to achieve by 2020 a share of smart meters 80 % of the total, after verification that the related long-term benefits outweigh the costs, i.e., cost–benefit analysis.[2] The implementation of intelligent metering systems designed to ensure long-term benefits to consumers. The third energy package does not set a specific goal for the implementation of intelligent metering systems in the gas sector, but the interpretative note on the retail markets indicates that it should be made within a reasonable time. A report of the European Commission in June 2014, "Benchmarking smart metering deployment in the EU-27 with a focus on electricity," (COM (2014) 356) has measured progress made in the implementation of smart meters in the EU 27 for the electricity sector; the analysis shows that considerable progress has been made. According to the report, as a result of positive cost–benefit analysis for the electricity in more than two-third of cases, member states are now committed to proceed with the introduction of smart meters (or have already completed). Nearly 45 million smart meters have already been installed in three member states (Finland, Italy, and Sweden), which represent 23 % of the installations planned in the EU by 2020. Below are the results of the cost benefit conducted by member countries with reference to the introduction of smart meters. In the electricity sector, 16 member states (Austria, Denmark, Estonia, Finland, France, Greece, Ireland, Italy, Luxembourg, Malta, the Netherlands, Poland, Romania, Spain, Sweden, and the UK) proceed or have already introduced large-scale smart meters by 2020 or earlier. In Poland and Romania, the cost–benefit analysis has produced positive results, but they have not yet adopted official decisions for the introduction. In the other EU countries, the cost–benefit analysis has failed or has yielded no effect. However, in the majority of EU countries, there is legislation in the field of smart meters for electricity, which establishes a legal framework for the introduction and its discipline specific issues such as the timetable for introduction, or fixed specifications for meters, etc. As regards the gas sector, only five member states (Ireland, Italy, Luxembourg, the Netherlands, and the UK) have decided to introduce smart meters by 2020 or earlier. Available data reported in the COM (2014) 356 indicate that a system of smart metering could cost an average of € 200 to € 250 per consumer and the smart metering systems should bring overall benefits to consumers amounted to € 160 for gas and € 309 for electricity, accompanied by energy savings of 3 %. The most challenging feature to be implemented refers to the frequency with which consumption data can be updated and made available to consumers and third parties acting on their behalf. This feature will help to provide consumers with information on the direct costs, which enables them to make informed choices about their consumption patterns and facilitates the development of new services and retail products.

[2] The approach recognizes that the impact of smart grid projects goes beyond what can be captured in monetary terms, and then it integrates the economic analysis with a qualitative analysis approach that includes a nonmonetary approach in terms of externalities such as impacts social (Joint Research Centre, European Commission, 2012).

The infrastructure for electricity generation and network investments has long-term returns and therefore requires a stable framework. To achieve EU and National Energy Policy objectives, a new global approach is needed in the generation, transmission, distribution, metering, supply, and consumption of energy. It is expected that the smart grids offer great benefits to all stakeholders of the electricity system. The grid operators can manage the network more efficiently and consumers can switch from passive receiver of electricity to active participants in the supply channel.

The projects of smart grids are not uniformly distributed in the EU. Most are in the 15 EU member states, while other countries are still lagging behind. In addition, most of the projects are concentrated in a few countries: Denmark, Germany, Spain, and the UK together account about half of the total number of projects in Europe. However, in almost all EU countries, a significant amount of investment has been allocated to projects that aim to integrate various smart grid technologies and applications. Most of the technology is unknown, but their integration is the current challenge.

4.4 Conclusions

Although the environmental benefits of reducing dependence on fossil fuels are clear, the storage of large quantities of energy from RES and the small-scale production still represent a major challenge to the stability and availability of electricity. The main difficulty stems from the intermittent nature of RES. Although hydroelectric sources offer a highly predictable supply of electricity, the availability of most of the RES is variable. The production of energy in wind farms is characterized by periods of high productivity followed by moments of stasis, while the performance of photovoltaic systems is less in case of cloudy weather or at night. Research and development activities go on and currently new methods are available to ensure the storage of energy generated from RES with particular reference to the field of solar thermal power that with new technology is able to operate even in the absence of direct light, with cloudy sky, and at night. Every step forward toward technology solutions that reduce the impact on the environment is of great importance, especially for what concerns the field of RES.

The addition of large quantities of RES undoubtedly requires more investment in smart technologies that allow for containment and a rapid and effective correction of problems during delivery. Smart grids make it possible to integrate RES with other electricity sources. Smart grids are bound to become increasingly important given the increasing reliance of the world economies by electricity. Another challenge related to RES concerns the location. Those in large scale, offshore, or in the middle of the desert are far from centers of demand, while small producers are often located in residential areas or in low industrially exploited areas, in which the local distribution network is not utilized for the reception and supply of electricity.

Compared with highly predictable performance of conventional power plants, of which most can be built in the vicinity of the communities served, those fueled by RES offer some inconvenience in most. However, most of the technologies necessary to overcome obstacles related to the use of RES are already available, while others are currently under study and development. Hand in hand with the development of new systems based on RES, the network evolves in order to provide a reliable electricity supply in line with demand.

4.5 Appendix: Projections for RES Electricity in 2020 in NREAP's EU Countries

(Source: EREC 2011)

Austria

RES 2020 projections	MW installed	RES electricity generation (GWh)	% in electricity consumption
Large hydro	7707	36,071	48.6
Hydro	1291	6041	8.1
Pumping	4285	2732	3.7
Geothermal	1	2	0.0
Photovoltaic	322	306	0.4
Wind onshore	2578	4811	6.5
Biomass	1179	4566	6.2
Biogas	102	581	0.8
Total	13,180	52,378	70.6[a]

[a] Pumping is not included

Belgium

RES 2020 projections	MW installed	RES electricity generation (GWh)	% in electricity consumption
Hydro	140	440	0.4
Geothermal	3.5	29.1	0.0
Photovoltaic	1340	1139	1.0
Wind onshore	4320	10,474	9.5
Biomass	2024.5	9599.4	8.7
Biogas	427	1439.1	1.3
Total	8255	23,121	20.9

Bulgaria

RES 2020 projections	MW installed	RES electricity generation (GWh)	% in electricity consumption
Hydropower	2549	3951	10.8
Photovoltaic	303	454	1.2
Wind onshore	1256	2260	6.2
Biomass	93	514	1.4
Biogas	65	357	1.0
Total	4266	7536	20.6

Cyprus

RES 2020 projections	MW installed	RES electricity generation (GWh)	% in electricity consumption
Photovoltaic	192	309	4.0
Solar thermal electricity	75	224	3.0
Wind onshore	300	499	7.0
Biomass	17	143	2.0
Total	584	1175	16.0

Czech Republic

RES 2020 projections	MW installed	RES electricity generation (GWh)	% in electricity consumption
Hydro	1125	2274	3
Geothermal	4.4	18	0
Photovoltaic	1695	1726	2
Wind onshore	743	1496	2
Biomass	–	3294.0	4
Biogas	417	2871.0	3
Total	3,984.4	11,679	14

Denmark

RES 2020 projections	MW installed	RES electricity generation (GWh)	% in electricity consumption
Hydro	10	31	0.1
Photovoltaic	6	4	0.0
Wind onshore	2621	6391	16.9
Wind offshore	1339	5322	14.1
Biomass	2430	6353	16.8
Biogas	349	2493	6.6
Total	6755	20,594	54.5[a]

[a] This figure differs from the one calculated in the Danish NREAP

Estonia

RES 2020 projections	MW installed	RES electricity generation (GWh)	% in electricity consumption
Hydro	7.8	30	0.3
Wind onshore	400	974	8.9
Wind offshore	250	563	5.2
Biomass	–	346	3.2
Total	658	1913	17.6

Finland

RES 2020 projections	MW installed	RES electricity generation (GWh)	% in electricity consumption
Hydro	3100	14,420	14.2
Wind onshore	2500	6090	6.0
Biomass	2920	12,640	12.4
Biogas	–	270	0.3
Total	8530	33,420	32.9

France

RES 2020 projections	MW installed	RES electricity generation (GWh)	% in electricity consumption
Hydro	23,496	71,702	13.1
Geothermal	80	475	0.1
Photovoltaic	4860	5913	1.1
Solar thermal electricity	540	972	0.2
Wind onshore	19,000	39,900	7.3
Wind offshore	6000	18,000	3.3
Tidal, wave, ocean	380	1150	0.2
Biomass	2382	13,470	2.5
Biogas	625	3701	0.7
Total	57,363	155,283	28.5[a]

[a] This figure differs from the percentage indicated in the French NREAP

Germany

RES 2020 projections	MW installed	RES electricity generation (GWh)	% in electricity consumption
Hydropower	4309	20,000	3.6
Geothermal	298	1654	0.3
Photovoltaic	51,753	41,389	7.4
Wind onshore	35,750	72,664	12.9
Wind offshore	10,000	31,771	5.7
Biomass	5029	26,019	4.6
Biogas	3796	23,438	4.2
Total	110,935	216,935	38.6

Greece

RES 2020 projections	MW installed	RES electricity generation (GWh)	% in electricity consumption
Hydro	4531	4873	7.1
Geothermal	120	736	1.1
Photovoltaic	2200	2891	4.2
Solar thermal electricity	250	714	1.0
Wind onshore	7200	16,125	23.6
Wind offshore	300	672	1.0
Biomass	40	364	0.5
Biogas	210	895	1.3
Total	14,851	27,270	39.8

Hungary

RES 2020 projections	MW installed	RES electricity generation (GWh)	% in electricity consumption
Hydro	66	237	0.5
Geothermal	57	410	0.8
Photovoltaic	63	81	0.2
Wind onshore	750	1545	3.0
Biomass	500	2688	5.2
Biogas	100	636	1.2
Total	1537	5597	10.9

Ireland

RES 2020 projections	MW installed	RES electricity generation (GWh)	% in electricity consumption
Hydro	284	923	2.8
Geothermal	5	35	0.1
Photovoltaic	5	13	0.0
Tidal, wave, ocean	500	1533	4.7
Wind onshore	4737	12,449	38.1
Wind offshore	2408	7383	22.6
Biomass	400	2453	7.5
Total	8339	24,789	75.8

Italy

RES 2020 projections	MW installed	RES electricity generation (GWh)	% in electricity consumption
Hydro	17,800	42,000	11.2
Geothermal	920	736	1.1
Photovoltaic	8000	9650	2.6

RES 2020 projections	MW installed	RES electricity generation (GWh)	% in electricity consumption
Solar thermal electricity	600	1700	0.5
Tidal, wave, ocean	600	1700	0.5
Wind onshore	12,000	18,000	4.8
Wind offshore	680	2000	0.5
Biomass	2620	12,760	3.4
Biogas	1200	6020	1.6
Total	43,823	98,885	26.4

Latvia

RES 2020 projections	MW installed	RES electricity generation (GWh)	% in electricity consumption
Hydro	1550	3051	35.2
Photovoltaic	2	4	0.0
Wind onshore	236	519	6.0
Wind offshore	180	391	4.5
Biomass	108	642	7.4
Biogas	92	584	6.7
Total	2168	5191	59.8

Lithuania

RES 2020 projections	MW installed	RES electricity generation (GWh)	% in electricity consumption
Hydro	141	470	3.4
Photovoltaic	10	15	0.1
Wind onshore	500	1250	9.0
Biomass	162	810	5.8
Biogas	62	413	3.0
Total	874.8	2958	21.3

Luxembourg

RES 2020 projections	MW installed	RES electricity generation (GWh)	% in electricity consumption
Hydro	44	124	1.9
Photovoltaic	113	84	1.3
Wind onshore	131	239	3.6
Biomass	30	190	2.8
Biogas	29	144	2.2
Total	347	781	11.8

Malta

RES 2020 projections	MW installed	RES electricity generation (GWh)	% in electricity consumption
Photovoltaic	27.8	42.7	1.4
Wind onshore	14.58	38	1.2
Wind offshore	95	216	6.9
Biomass	15.18	86	2.7
Biogas	7.34	50	1.6
Total	159.9	432.7	13.8

The Netherlands

RES 2020 projections	MW installed	RES electricity generation (GWh)	% in electricity consumption
Hydro	203	714	1.0
Photovoltaic	722	570	0.0
Wind onshore	6000	13,372	10.0
Wind offshore	5178	19,036	14.0
Biomass	2253	11,975	9.0
Biogas	639	4664	3.0
Total	14,995	50,331	37.0

Poland

RES 2020 projections	MW installed	RES electricity generation (GWh)	% in electricity consumption
Hydro	1152	2969	1.7
Photovoltaic	3	3	0.0
Wind onshore	6150	13,710	8.1
Wind offshore	500	1500	0.9
Biomass	1550	10,200	6.0
Biogas	980	10,200	2.4
Total	10,335	32,400	19.1

Portugal

RES 2020 projections	MW installed	RES electricity generation (GWh)	% in electricity consumption
Hydro	9548	14,073	21.8
Geothermal	75	488	2.3
Photovoltaic	1000	1475	2.3
Solar thermal electricity	500	1000	1.5
Tidal, wave, ocean	250	437	0.7
Wind onshore	6800	14,416	22.4
Wind offshore	75	180	0.3

RES 2020 projections	MW installed	RES electricity generation (GWh)	% in electricity consumption
Biomass	802	2991	4.6
Biogas	150	525	0.8
Total	19,200	35,588	55.2

Romania

RES 2020 projections	MW installed	RES electricity generation (GWh)	% in electricity consumption
Hydro	7729	19,769	26.9
Photovoltaic	260	320	0.4
Wind onshore	4000	8400	11.4
Biomass	405	1950	2.6
Biogas	195	950	1.3
Total	12,589	31,389	42.6

Slovakia

RES 2020 projections	MW installed	RES electricity generation (GWh)	% in electricity consumption
Hydro	1812	5400	16.2
Geothermal	4	30	0.0
Photovoltaic	300	300	0.9
Wind onshore	350	560	1.7
Biomass	170	850	2.6
Biogas	110	860	2.6
Total	2746	8000	24.0

Slovenia

RES 2020 projections	MW installed	RES electricity generation (GWh)	% in electricity consumption
Hydro	1353	5121	32.9
Photovoltaic	139	139	0.9
Wind offshore	106	191	1.2
Biomass	34	309	2.0
Biogas	61	367	2.3
Total	1693	6127	39.3

Spain

RES 2020 projections	MW installed	RES electricity generation (GWh)	% in electricity consumption
Hydro	22,362	39,594	10.6
Geothermal	50	300	0.1
Photovoltaic	8367	14,316	3.8
Solar thermal electricity	5079	15,353	4.1
Tidal, wave, ocean	100	220	0.1
Wind onshore	35,000	70,502	18.8
Wind offshore	3000	7753	2.1
Biomass	1187	7400	2.0
Biogas	400	2617	0.7
Total	69,845	150,032	40.2

Sweden

RES 2020 projections	MW installed	RES electricity generation (GWh)	% in electricity consumption
Hydro	16,326	67,929	44.0
Photovoltaic	8	4	0.0
Wind onshore	4365	12,000	7.8
Wind offshore	182	500	0.3
Biomass	2872	16,700	10.8
Biogas	42	53	0.0
Total	23,786	97,186	62.9

The UK

RES 2020 projections	MW installed	RES electricity generation (GWh)	% in electricity consumption
Hydro	2130	6360	1.7
Photovoltaic	2680	2240	0.6
Tidal, wave, ocean	1300	3950	1.0
Wind onshore	14,890	34,150	9.1
Wind offshore	12,990	44,120	11.7
Biomass	4240	26,160	6.9
Total	38,230	116,980	31.0

References

Archer CL, Caldeira K (2009) Global assessment of high-altitude wind power. Energies 2(2):307–319

Canale M, Fagiano L, Milanese M (2009) KiteGen: A revolution in wind energy generation. Energy 34(3):355–361

Edison Electric Institute (2013) Straight talk about net metering, U.S.

EurObserv'ER (2014) http://www.eurobserv-er.org. Accessed 14 July 2014 European Commission (2014) Communication (2014) 356 REPORT FROM THE COMMISSION Benchmarking smart metering deployment in the EU-27 with a focus on electricity

European Commission (2009a) Directive 2009/28/EC of the European Parliament and of the Council of 23 April 2009 on the promotion of the use of energy from renewable sources and amending and subsequently repealing Directives 2001/77/EC and 2003/30/EC. Off J Eur Union L 140:16–62

European Commission (2009b) Directive 2009/72/EC of the European Parliament and of the Council of 13 July 2009 concerning common rules for the internal market in electricity and repealing Directive 2003/54/EC. Off J Eur Union L 211:55–93

European Commission (2009c) Directive 2009/73/EC of the European Parliament and of the Council of 13 July 2009 concerning common rules for the internal market in natural gas and repealing Directive 2003/55/EC. Off J Eur Union L 211:94–136

European Renewable Energy Council (EREC) (2011) Mapping renewables energy pathways towards 2020. Belgium

European Wind Energy Association (EWEA) (2014) Wind in power—2013 European statistics. Brussels

Giordano V, Onyeji I, Fulli G, Jiménez MS, Filiou C (2012) Guidelines for conducting a cost-benefit analysis of smart grids projects. Reference Report by the Joint Research Center of the European Commission

Chapter 5
EU Scenarios for a Sustainable RES Strategy

5.1 Introduction

The development of renewable energy source (RES) fits in a global context of increasing attention to climate change and energy security issues. Recent analyses of energy scenarios (European Commission 2009; European Commission 2014; IEA 2014a) show the possibility for the global energy system of a transition on the long term towards a model of production and consumption increasingly based on RES. In this chapter, we propose our RES development scenario for the EU in order to achieve RES amount by 2030 at lower costs than those expected by the European Commission. Consistent with economic theory, our results suggest that EU policy makers should follow purely a logic of cost minimization for the EU as a whole, according to which EU countries with lower prices (levelized cost of energy (LCOE)[1]) of production should produce more and vice versa. Indeed, it is conceivable that this could spur some opportunistic behavior: countries with higher costs would try to reduce their allocation but this will increase the cost for the whole EU; so they can try to negotiate some compensation. The decoupling between RES and cost share at the country level leads us to strongly affirm that cooperation among EU countries is essential to achieve the sustainable development of RES. To this aim, it is appropriate to devise a radically new energy and environmental policy. Given that the objectives are challenging at a global level, even EU policies should be equally challenging. In other words, it is time to conceive the EU as a single "Nation" facing the environmental issue and indeed, it is necessary to have a unique European policy supporting RES production. This implies to overcome the fragmentation of 28 local policies. According to most of the literature, we advocate price driven instruments rather than quantity driven instruments to support RES development. Such policies leave each country free to choose the suitable mix of RES. It should be kept in mind that some RES are disadvantageous compared to other RES, given the differences in the natural endowments of countries. This implies that the RES should be produced

[1] See Chap. 3, Sect. 3.3 of this book.

S. Bigerna et al., *The Sustainability of Renewable Energy in Europe*,
DOI 10.1007/978-3-319-12343-1_5

and exploited where it is most convenient: wind power where there is a lot of wind and photovoltaic, and solar thermal where there is a lot of sun.

Finally, we critically assess the implications of our analysis, pointing out that our analysis is mostly technical and deliberately refrains from considering some exogenous factors that influence the development of RES in the medium and long term.

The chapter proceeds as follows: Sect. 5.1 provides a description of RES scenario according to the International Energy Agency (IEA) (2014a) at global and European levels. Section 5.2 describes RES scenario of the EU energy and environmental policies according to the 2030 framework for environmental policies. Furthermore, we provide a cost–benefit analysis for the EU countries. Section 5.3 provides our proposal for a new policy strategy at the EU level. Section 5.4 concludes the chapter.

5.2 World Energy Outlook Scenarios for RES

The global consensus on environmental policies expects in the coming years an important deployment of RES in order to fight climate change. Energy from RES can become the energy basis of the future and the world's energy debate has led to outline a number of energy scenarios worldwide.

In order to provide a brief summary of the development scenarios for energy, with particular reference to RES, we take as reference the scenarios outlined by the IEA (2014a) in the World Energy Outlook-2014 (WEO-2014). It shows once more that the pursuit of "climate" requires the development and diffusion of RES far greater than those that current policies can ensure. In particular, there are three different proposed scenarios for the evolution of the RES' market. First, the new policies scenario, which is the central scenario of the WEO-2014: it incorporates policies and measures in place relating to energy markets and those which have been adopted until mid-2014; it also takes into account the credible commitments that have been announced. Second, the current policies scenario, which takes into account the energy policies in force until the middle of 2014. Therefore, it describes a scenario in which governments do not implement policies currently under discussion nor new ones are introduced. The objective of the current policies scenario is to describe how energy markets evolve if the trends of energy demand and supply remain unchanged. It can be seen as a kind of business-as-usual scenario since it describes the consequences of inaction and allows to assess the effectiveness of recent implemented environmental and energy policies. Third, the 450 scenario, expects to achieve the target to limit the rise in global atmospheric temperature by 2 °C by 2050, thanks to the reduction in the concentration of GHG at around 450 ppm. According to the 450 scenario, it is expected to reach a peak in the concentration of greenhouse gas (GHG) in the atmosphere in mid-2050 to a level slightly higher than 450 ppm, while around 2100 the concentration of GHG will stabilize at 450 ppm. For the period up to 2020, it is assumed to undertake the actions decided in the Cancun Agreements. After 2020, it is assumed that Organization for Economic

Co-operation and Development (OECD) countries implement measures to reduce emissions consistent with the objective of 450 scenario and invest about US$ 100 billion to reduce emissions in non-OECD countries.

Consequently, it is important to the development of RES that should increasingly become part of the global energy mix, especially for the power sector. The growth rates observed so far suggest that RES are expected to grow strongly again until 2040 that is the period considered in the WEO-2014, by holding the support measures for RES in place. In the USA, the market for RES has grown strongly thanks to RES support policies, such as subsidies that are given up to 30% of the investment costs for projects related to RES. In Europe, there has been a rapid expansion of RES, especially wind and solar, linked to the requirements of the EU RES directive (Directive 2009/28/EC) and consequently of national targets. The growth of energy demand and concerns over energy security and pollution suggest that even in non-OECD countries the development of RES is intended to widen.

We present a brief summary of the three scenarios in order to give a quantitative outlook of RES use by type and scenario worldwide. In all three scenarios, it is provided for an increase in the use of RES, mainly because energy policies in place aimed at making the RES technologies competitive with fossil fuel-based technologies (Table 5.1).

We notice the 450 scenario envisages a share of RES in the total demand for primary energy sources of 30% in 2040, compared to 19% of new policies scenario, doubling the current share of 13% in the current policies scenario. However, this perspective is possible not only with an increase in "modern RES," including hydropower, wind, solar, geothermal, marine, and bioenergy, but also through the gradual replacement of the traditional use of biomass with other forms of energy production with lower environmental impact especially in developing countries.

In the new policies scenario, the future of RES is outlined in the light of specific issues such as the integration of intermittent RES in particular in the electrical systems, the implications for the markets of the development of RES and cost competitiveness of RES compared to fossil fuels. The new policies scenario is quite similar in the current policies scenario 2020 in terms of share of RES in electricity generation (26% of the new policies scenario versus 25% of the current policies scenario), while the new policies scenario becomes more effective in 2040 (33% of the new policies scenario versus 25% of the current policies scenario). Given the similarity of the two scenarios, we skip the discussion of the current policies scenario which is more fainthearted with respect to the new policies scenario in terms of RES development to 2040.

On the contrary, the 450 scenario's perspective takes a completely different scale. Although it is difficult to assess the plausibility of the 450 scenario, we deem it is difficult to achieve its objectives.

In fact, in the new policies scenario, RES electricity generation share should reach the 33% of the total electricity generation and for the two main intermittent sources addressed, i.e., wind and solar, in the new policies scenario it is expected that the installed capacity quadruples in the case of wind farms and increases seven times in the case of solar panels by 2040, reaching a share of global electricity

Table 5.1 World RES use by type and scenario. (Source: IEA 2014a)

		New policies scenario		Current policies scenario		450 scenario	
	2012	2020	2040	2020	2040	2020	2040
Primary demand (Mtoe)	*1802.0*	*2254.0*	*3455.0*	*2223.0*	*3095.0*	*2276.0*	*4658.0*
USA	136.0	186.0	346.0	181.0	290.0	188.0	555.0
European Union	198.0	263.0	390.0	255.0	351.0	267.0	504.0
China	316.0	410.0	589.0	397.0	501.0	414.0	842.0
Share of global total primary energy demand	13%	15%	19%	15%	15%	16%	30%
Electricity generation (TWh)	*4807.0*	*7263.0*	*13,229.0*	*7010.0*	*11,046.0*	*7329.0*	*17,973.0*
Bioenergy	442.0	764.0	1569.0	740.0	1299.0	768.0	2261.0
Hydro	3672.0	4553.0	6222.0	4458.0	5862.0	4561.0	6943.0
Wind	521.0	1333.0	3345.0	1254.0	2552.0	1376.0	4953.0
Geothermal	70.0	120.0	378.0	113.0	287.0	121.0	557.0
Solar photovoltaic	97.0	449.0	1 291	408.0	832.0	459.0	1 982
Concentrating solar power	5.0	41.0	357.0	34.0	173.0	42.0	1 158
Marina	1.0	3.0	66.0	3.0	41.0	3.0	119.0
Share of total generation	21%	26%	33%	25%	25%	27%	51%
Heat demand[a]	*345*	*431*	*716*	*431*	*670*	*450*	*932*
Industry	198	242	367	246	381	249	447
Buildings and agriculture	147	189	348	185	289	201	485
Share of total final demand	10%	11%	16%	11%	14%	12%	23%
Biofuels (mboe/d[b]*)*	*1.3*	*2.2*	*4.6*	*1.8*	*3.6*	*2.1*	*8.7*
Road transport	1.3	2.2	4.5	1.8	3.6	2.1	7.2
Aviation	–	–	0.0	–	0.0	–	1.5
Share of total transport	2%	4%	6%	3%	5%	4%	20%
Traditional biomass (Mtoe)	*758*	*755*	*648*	*760*	*671*	*750*	*633*
Share of total bioenergy	56%	49%	32%	49%	35%	48%	25%
Share of RES energy demand	42%	34%	19%	34%	22%	33%	14%

[a] excludes traditional biomass

[b] million barrels of oil equivalent per day

generation by 6 % for wind power and 3 % for solar photovoltaic. This rapid growth is expected to be localized mainly in the emerging economies, India and China, although the EU, the USA, and Japan should continue to play an important role in RES' markets. In the face of significant developments identified by this scenario, significant implications are highlighted in terms of adaptation, new design and investment in energy grids for the exploitation of intermittent sources and in terms of the need for capacity of traditional reserves, that are flexible plants for dispatching energy to cope with the demand for energy that might not be covered by RES.

The 450 scenario is even more challenging since it requires practically a revolution that would involve a radical afterthought, even institutional, of the power systems since the share of the RES electricity generation in 2040 that could reach 51 % of the total, with solar and wind power installed capacity equal to 50 % of peak demand global. This is difficult to implement in light of the impact of RES on liberalized electricity markets, in which the projected future low marginal costs of production of energy from RES, due to the effect of the merit order, tend to reduce the price of electricity and displace the traditional plants shrinking profit's margins, with the risk of not covering investment cost.

In summary, we provide estimates of investment needed to get the objectives outlined in both the new policies scenario and 450 scenario (IEA 2014b). Estimates in the new policies scenario provide for the need for cumulative investment in energy of US$ 48 trillion between 2014 and 2035, of which US$ 40 trillion in energy supply and US$ 8 trillion in energy efficiency. The US$ 40 trillion in energy supply is split into US$ 23 trillion in fossil fuel extraction, US$ 6 trillion in RES, US$ 1 in nuclear, and US$ 7 trillion in transmission and distribution. Most of the investments, i.e., 63 % of energy supply investment, has to be made in non-OECD and the remaining US$ 14.5 trillion in OECD countries. Instead, with regard to energy efficiency, investments are higher for OECD countries, at around US$ 4.63 trillion, than non-OECD countries (US$ 3.14 trillion), reflecting the size of OECD energy markets.

By focusing on the power sector, global investment forecasted amounts to US$ 16.4 trillion between 2014 and 2035, of which US$ 9.6 trillion for power plants and US$ 6.8 trillion for transmission and distribution infrastructures. Investments in power plants are mainly aimed at non-hydro RES, mostly wind and solar photovoltaic, because their capacity is projected to increase over time, in spite of fossil fuels.

The achievement of the objectives outlined in the 450 scenario implies cumulative investments to 2035 of US$ 53 trillion of which US$ 13.5 trillion in energy efficiency and $ 39.4 trillion in energy supply. Compared to the new policies scenario, the 450 scenario requires heavier efforts on energy efficiency to reduce pollution, so that the cumulative investment in the 450 scenario is greater than $ 5 trillion with respect to the new policies scenario. The energy supply requires investments mainly by non-OECD countries (US$ 24.1 trillion) and the remaining US$ 14.9 trillion by OECD countries.

A key point to achieve the 450 scenario's goals is the deployment of less mature low carbon technologies, as RES and carbon capture and storage. Power sector investments amount to US$ 19.3 trillion from 2014 to 2035, of which RES represents 70 % of total investments in 2035. RES investments in the 450 scenario are almost one fifth higher than in the new policies scenario. This is basically due to the fact that RES investments in the 450 scenario aim to develop even less mature RES technologies, as geothermal and marine energy. The remaining 30 % investments in the power sector ($ 5.9 trillion over 2014–2035) refer to transmission and distribution infrastructures.

5.3 RES Scenario in the EU 2030 Framework

The novelty of the EU energy and environmental policies is constituted by the 2030 framework for climate and energy policies (COM (2014)0015) presented by the European Commission in January 2014. The text of the new EU strategy is based on reducing emissions, RES implementation, and increasing energy efficiency. The most important and controversial issue of the 2030 framework is represented by the new EU target for the RES share in gross energy consumption in 2030. The European Commission proposes an RES target of 27 % in gross energy consumption which is binding only at EU level but not for individual EU member countries; it differs from the previous 2020 climate and energy package with binding RES target for each EU member countries. The new RES objective is a goal "to reach all together." The quantitative value of this objective is determined by the weight of the RES that is expected as a natural development to achieve the objective −40 % for GHG. As we will clarify further, the nonbinding RES share per EU country is a controversial issue since it requires much more need for cooperation among member states.

Furthermore, the communication 2014 (0015) proposes a mandatory reduction of GHG emissions by 40 % compared to the emission levels in 1990. The target is divided between −43 % (compared to 2005 emissions) reduction at the European level to be distributed among economic sectors included in the EU emission trading system (EU ETS) and −30 % (compared to 2005 emissions) reduction at the European level to be distributed among the member states for the economic sectors not covered by the EU ETS, on the basis of indicators to take into account the distribution among countries and the sustainability of needed investments. The Commission also proposes that the ETS system is reformed, implementing a reserve for market stability operational in 2021, i.e., since the beginning of the next trading period. The mechanism of the reserve falls within the dynamics of supply and management systems and it should increase the responsiveness of the ETS to systemic shocks on the demand side or the supply side.

The role of energy efficiency in the context in 2030 is not identified by the European Commission in any specific target in terms of increasing energy efficiency by 2030; it will be further considered with a review of the energy efficiency directive ready by the end of 2014.

The text of the new EU strategy on reducing emissions, RES production and energy efficiency improvement are replacing the 2020 climate energy package which should have ended in 2020. Indeed, recent data demonstrate that the path of energy and environment that the EU has set itself for 2020 with the 2020 climate and energy package is progressing well: by the end of 2012 the EU countries overall have reduced emissions by 18 % on 1990 levels and also with reference to the RES, in 2011 around 13 % of the energy produced came from RES, a given upper 11.7 % expected for 2011–2012, a higher figure than 11.7 % expected for 2011–2012. We briefly recall that the crucial feature of the 2020 climate and energy package enacted in 2009 consists of binding measures for all EU member countries in terms of RES share to curb emissions of GHG and get to the progressive decarbonization of the EU.[2] It outlines the efforts distribution to which all EU member countries must undertake to ensure RES goals within the package.

In order to ensure a better chance of success at the policy outlined in the COM (2014)0015, the European Commission proposes the establishment of a regulatory framework of governance, based on national plans for competitive, secure, and sustainable energy. These plans, which will follow the guidelines that will be developed by the EU, must be prepared by individual member states through an iterative process involving at least the neighboring states and be assessed by the European Commission in terms of their ability to meet the energy and environmental EU objectives. The performance of the plans and the achievement of the objectives will be examined by specific indicators as energy price differentials with competitors outside the EU, diversification of supply sources, the development of smart grids and interconnections, the level of interaction of European markets, the degree of competition and concentration of national or regional energy markets, and technological development.

On the adequacy and appropriateness of the objectives proposed by the European Commission in the 2030 framework, there has been opened a wide debate at both national and European levels. On the one hand, the world of industry, particularly the manufacturing industry, has complained that the definition of new targets for the reduction of GHG could have an adverse effect on the price of energy in Europe, increasing the cost differential with respect to the energy cost of other market areas. In particular, it has been observed that in the USA also through the production of shale gas, the cost of energy has been considerably lowered. An increase in the cost of energy in Europe would therefore seriously harm the competitiveness of enterprises. Major utilities and energy operators have highlighted the need for a single target on the reduction of GHG to support a technology-neutral approach between the technologies with low CO_2 emissions. On the other hand, there are concerns among stakeholders about the lack of binding target for RES at national levels. The COM(2014)0015, avoiding to set a national target for the RES and in the absence of a real governance of coordination for the achievement of the EU goal fixed to 27 % of gross energy consumption from RES, would seal a reversal of direction compared to the previous 2020 climate and energy package, bringing instability and regulatory perspective of the RES sector.

[2] The 2020 climate and energy package, enacted in 2009, consists of a series of measures that mainly concern the 20 % reduction of at least 20 % of GHG emissions resulting from energy consumption in the EU compared to 1990 levels by 2020, the 20 % increase of energy produced from RES by 2020; the 20 % improvement in energy efficiency.

Table 5.2 Scenarios developed in the report of Enerdata's global energy forecasting department (2014) and IEA (2014a). (Source: our elaboration from Enerdata (2014) and IEA (2014a))

Scenario	GHG emissions (%)	RES share (%)	Energy efficiency target	Credits[a] (%)	Technological innovation
WEO					
a	20	19			
b	20	15			
c	30	30			
Enerdata					
d	40	30	30	0–5–10	x
e	50	-	-	5–10	
f	60			5	

a new policies scenario; *b* current policies scenario; *c* 450 scenario; *d* scenarios with 40% GHG emissions reduction; *e* scenarios with 50% GHG emissions reduction; *f* scenarios with 60% GHG emissions reduction

[a] Credits are available annually for the EU ETS sectors at 144 Mt CO_2 and for the non-ETS sectors at 138 Mt CO_2 in the case of 5%; in the case of 10% the percentages are doubled respectively to 288 Mt CO_2 and 276 Mt CO_2

However, the controversy of public opinion toward some policies of decarbonization of the EU seems to be suffering from a certain myopia. Whatever way, the road to a low carbon economy must be undertaken by the EU.

At this point, we think it is very important to analyze costs and benefits of the scenario proposed in COM 2014(0015) to arrive at a definite negotiating among EU countries.

Unfortunately, cost–benefit analyses of the scenario proposed in COM 2014(0015) have been long overlooked; there are not that many cross-country analyses. Currently, such studies are increasing; we propose one of the most recent studies on this regard (Enerdata's global energy forecasting department 2014). In particular, there are analyzed costs and benefits for all EU member countries under various scenarios for both type and level of goals considered in the package to 2030. These scenarios include GHG target increasingly stringent, from 40 to 60% compared to 1990 levels, with different levels of RES' share, access to international credits, and so on. As a reference, we also show the macro objectives defined in the main scenarios by the IEA (2014a) (Table 5.2).

By focusing on the Enerdata's scenarios, the potential 40% reduction of GHG emissions requires for EU member countries as a whole an abatement cost of € 30 billion in 2030 compared to the reference case (it includes the EU policies up to 2020), that is the EU 0.2% of gross domestic product (GDP) in 2030. It is shown that the relationship between abatement costs and GHG targets in nonlinear. Indeed, a 50% reduction of GHG emissions in 2030 compared to 1990 levels would cost to the EU € 95 billion in 2030 which is 0.5% of GDP in 2030. It is also noted that the power sector is the one that can mostly contribute to reducing GHG emissions. Indeed, RES have mostly seen their greatest development and implementation in the power sector. According to the reference case, the share of RES in 2030 would amount to 25 % of the energy final consumption. The RES should cover the power sector in relation to the gross final energy consumption by an amount equal to 68 %, transport for 17 %, and heating for 15 %. Under the scenario, 40 % target of GHG emissions with RES share of consumption at the EU level equal to 30 % in 2030 would decrease the share of coal in primary energy consumption dramatically in favor of RES and partially of nuclear industry. The cost of reducing 40 % GHG emissions with the target of 30 % RES in the gross final energy consumption increases by 36 % compared to the sole objective of GHG, that is equivalent to an extra 0.1 % of GDP (€ 12 billion). The report notes that this cost is mainly due to a mismatch between RES potential and EU member countries' goals considering a scenario that extends the burden sharing from 2020 to 2030.

5.4 Our Proposal for a New Policy Strategy

Given the target to 2030 of 27 % RES in final energy consumption proposed by the 2030 framework for climate and energy policies, effort in terms of cost to reach the overall target is different among EU countries as their starting point is different: all EU member states increased their share of RES in total consumption heterogeneously. It is a crucial issue for the EU to figure out the best strategy for allocating RES' objectives in terms of quantity to be reached at minimum costs and then to ensure participation. The relationship between costs of achieving RES objectives and the RES target is nonlinear. There exists costs' differences among EU countries given their relative price (LCOE) of production and then some of them would pay less, while other countries would have to pay more to reach the RES target. This issue is really crucial in addressing the true spirit of cooperation among EU members countries.

It is conceivable that countries with higher costs would try to reduce their RES allocation, but this will increase the cost for the whole EU. So they can try to negotiate some compensation. This is not extravagant, if all countries behave honestly. The strategies planned by governments imply different costs that may be prohibitive if the other EU countries do not make comparable efforts. Indeed, EU countries might try to behave as free riders not participating in EU RES agreement, thus imposing lower costs on the domestic economy at the expense of other countries.

For instance, Nordhaus (2009) highlights the problem of coordination among countries that undermines the achievement of the objectives of the 2020 climate and energy package. Every country wishes to stimulate new activities, new investment, more jobs in their own territory, using an appropriate mix of local taxation and subsidies, in collaboration with other instruments of command and control. However, countries may have an incentive to behave as a free rider, or to impose costs that are borne as little as possible on its economy, enjoying the benefits created by other countries (Barrett 1994). More generally, many environmental problems are concerned by the fact that when many countries are involved in international agreements, it is possible that some of them may find convenient to not participate (Kolstad 2000; Stavins 1995). The improvement of environmental quality takes the nature of an international public good, and as such, it poses problems of market failure that are well known to economists. The maximization of the overall net benefits clashing with private incentives of individuals raises the behavior of free riding, which in the presence of global environmental problems become even more attractive. Most of the literature on opportunistic behavior of environmental coalitions (Carraro et al. 2006; Rathman 2007) argues that a noncooperative outcome is reached when countries first decide whether or not to cooperate, and then set their environmental policy (Bollino and Micheli 2012). Participation in environmental agreements is linked to earnings that nations can get: climate change is a global problem and in the fight against climate change, every nation believes to undertake efforts that are much greater than the benefits. Barrett (1994) notes that the failure to participate in environmental agreements is due to lack of incentives. In particular, he finds that positive incentives, which are market-based instruments, can reduce overall costs but are less effective in inducing participation in the agreement. Negative incentives, such as the reciprocal measures, the financial penalties, and trade restrictions are necessary along with positive incentives to stimulate cooperation.

The most difficult challenge to international cooperation already verified empirically is to induce global participation agreement. International agreements on the environment are the appropriate response to the need for intervention for the problems related to climate change. The 2020 climate and energy package is an example of a mandatory distribution of the efforts to which all EU member countries must undertake to ensure that the EU meets the objectives of reducing emissions by 2020.

With reference to the 2030 framework for climate and energy policies, the new directive will not contain any mandatory target for individual nations in terms of RES, but only a target to achieve together. With the view to minimizing overall costs to achieve the goal of RES quantities to 2030, in this concluding part of the book, we propose a new policy tool allowing EU policy makers to assess the likelihood that countries behave as free riders in relation to RES' goals.

The technological developments are subject to a whole range of uncertainties in the long run. In this respect, it is quite difficult to forecast the rate of diffusion and the intensity of cost reduction, i.e., the increase in productivity, which is resulting from technological developments. On the one hand, the usual assumption is making explicitly reference to the concept of learning by doing, which provides a quantitative measure of the speed of cost reduction through time, but it is intrinsically based on exogenous events. On the other hand, assuming endogenous deployment of new technologies is a rather arbitrary exercise when considering such a long-term horizon, like the 2030 horizon of the 2030 framework for climate and energy policies.

In any case, different assumptions on future technology costs have important effects in shaping the final result, especially considering that these effects are cumulated through a very long period of time. However, assuming a uniform diffusion among countries, would yield numerical results that are quite homothetic.

For instance, if we consider a technological breakthrough in the wind power sector, it results in an equiproportional cost reduction of 65 % in 2030 compared to 2011 levels for all EU countries. By taking into account the effect of learning by doing inertia that does not consider such a technological breakthrough, costs' reductions in 2030 amount to 11 % compared to 2011 levels. The same can be said for solar thermal sector: a technological breakthrough might bring a projected costs' reduction of electricity generation from solar thermodynamic in 2030 by 68 % compared to 2011 levels, while the inertial learning by doing should lower costs by 23 % in 2030[3].

We have designed two alternative scenarios up to 2030 for the introduction of several technological breakthroughs, differentiated by country, leading to a quantitative measure of projected costs' reduction of electricity generation from the main three RES: solar and thermal (CSP) photovoltaic, wind energy, and biomass. Operationally, in the two scenarios, we have labeled ENESH and RESSH, we assume that the cost reduction in each EU country is proportional to the relative share of that country in the total EU energy consumption and RES deployment in 2011, respectively. So the maximum reduction cost takes place only in the leader country while all other countries will have costs reduction lower than the one that occurs in the leader country. The rationale is as follows. In the ENESH scenario, we assume that the success of deployment of new technologies is a function of the size of the energy sector in 2011 in each country. This is based on the concept of economies of scales which impacts on the technology innovation; therefore, the larger is the energy sector, the higher is the probability (or the lower the associated uncertainty) of success of the country's effort in research and development of new RES technologies. We assume a metric that yields a given cost reduction for each country, within the range of minimum and maximum values, as reported in the literature, for each RES, i.e., photovoltaic, wind, and biomass, according to the ranking of the energy consumption share in the EU in 2011. In this way, it follows that larger countries have a comparative cost advantage in 2030, because they would experience a larger cost reduction than smaller countries.

In the RESSH scenario, we assume that successful cost reduction is a function of the size of the RES deployment in 2011 in each country. The idea is that historical vocation somehow impacts on future success of RES innovation. Therefore, the larger the RES deployment in the initial period, the higher the success in reaching further technological developments in the future. Like the previous case, we assume that cost reduction in each country is proportional to the ranking of the RES deployment share in the EU in 2011. In this way, it follows that countries with larger RES sectors would have a larger cost reduction in 2030 than those lagging behind in RES deployment today.

Based on the estimation of convex cost functions for each technology in the EU regions, we have computed the additional cost that all other countries would have to incur, if some country behaves as a potential free rider, because she avoids complying with the targets of climate and energy package. This additional cost arises

[3] See Chap. 4, Sect. 4.3 of this book.

Table 5.3 Estimated penalty of limited participation in EU climate change policy in 2020 and 2030. (Source: Bollino and Micheli (2014); Eurostat (2014))

Nonparticipating country	Penalty 2020	Penalty 2030 (ENESH scenario)	Penalty 2030 with tech change (RESSH scenario)
Austria	1.74	1.73	1.69
Belgium	3.85	3.83	3.75
Bulgaria	0.44	0.44	0.46
Cyprus	0.12	0.12	0.12
Czech Republic	1.11	1.10	1.06
Denmark	3.07	3.06	3.07
Estonia	0.28	0.27	0.29
Finland	4.04	4.02	3.80
France	12.21	12.15	12.70
Germany	29.47	28.98	30.55
Greece	2.50	2.48	2.71
Hungary	0.84	0.84	0.80
Ireland	3.18	3.18	3.38
Italy	6.73	6.65	6.68
Latvia	0.27	0.26	0.26
Lithuania	0.35	0.35	0.35
Luxembourg	0.08	0.08	0.08
Malta	0.06	0.05	0.06
The Netherlands	7.48	7.46	7.61
Poland	4.55	4.54	4.50
Portugal	2.72	2.71	2.84
Romania	1.56	1.56	1.62
Slovakia	0.29	0.29	0.28
Slovenia	0.11	0.11	0.10
Spain	14.74	16.64	15.78
Sweden	6.08	6.05	5.80
The UK	19.45	19.39	19.86

because there are increasing marginal costs for the remaining countries which yields a more than proportionate increase in their burden (Table 5.3)

Given RES quotas established by the 2020 climate and energy package in relation to the RES directive (2009/28/EC), we estimate the minimum costs to reach the overall target of RES quantity, given their relative price (LCOE) of production. In particular, we calculate that the cost falls on the community if some countries do not participate in the achievement of fulfilling the quota imposed by the RES directive 2009/28/EC (European Commission 2009). Incomplete participation raises abatement costs for remaining countries by a penalty factor which depends on the convexity of the abatement's marginal cost. Consequences of defection by EU countries within the 2020 climate and energy package is described in the second column of Table 5.1: for instance, if Spain does not participate to the 2020 climate package, it would impose an additional 14.74 % to costs of the complying group. If we suppose Italy does not make effort to reach the RES target to 2020, the cost penalty associated to this partial participation would lead to additional cost of 6.73 % for the other participating countries. The additional costs Germany imposes on EU participating countries by noncomplying the RES target would amount to 29.47 %. If France does not make efforts to reach the RES target, it would cost an additional 12.21 % to the EU as a whole. Values are different in case of the EU12 member countries. Indeed, there is still a strong relationship between income and polluting emissions: a country with large GDP tends to increase global emissions, unless the increase in income stimulates a reduction in GHG intensity. The relative increase in GHG emissions calls for RES production in order to meet the goals of environmental policy. If we suppose Poland does not make efforts to reach the RES target to 2020, the cost penalty associated to partial participation would cost an additional 4.55 % on EU participating countries. For instance, by considering Bulgaria and the Czech Republic, the relative percentages amount respectively to 0.44 and 1.11 %.

Different levels of penalty factors are obtained in the cases of ENESH and RESSH scenarios. Our estimates, extended to 2030, show that the penalty factor is basically lower in the ENESH scenario with respect to RESSH scenario, meaning that defection is less expensive for the EU as a whole in the case in which the future cost reduction (as explained above) of each country is taken according to the relative share of the country in the total EU energy consumption rather than the share of RES deployment. Let us take a numerical example. The penalty associated to the nonparticipation of Spain in the 2030 framework would impose additional costs on the remaining EU countries for an amount of 14.64 % in the ENESH scenario versus 15.78 % in the RESSH scenario. Then, different levels of penalty factors in all the three cases described mean that there are different abatement costs across EU member states.

It is therefore necessary a cooperative scenario based on the minimization of the overall costs for the EU, even if this involves the implementation of compensation mechanisms. A similar mechanism would also be a tool to revitalize the EU ETS market. Now more than ever there is a need for EU countries to feel united, overlooking immediate personal interests. The key idea of our scenario is that the EU appears as a single entity at the tables of global negotiations on environmental policies. This provides that the RES shares are disconnected from the interests of each country and therefore RES shares purely follow a logic of cost minimization of the EU. If this policy will lead to unequal concentrations of RES production in the territories because of environmental vocations of each country, then there should be a fair mechanism for compensation among EU countries. This may be the way to resume the leadership that was the spirit of the 2020 climate and energy package through which the EU intended to act as a leading example in the world.

We think it is possible to implement a procedure of minimization of the overall costs for the EU. Indeed, we have simulated a social welfare function to minimize the EU overall costs for a given amount of total RES target at European level. In

Table 5.4 Social welfare function simulation: photovoltaics, target year 2020. (Source: Our elaboration from Eurostat (2014))

Country	Target quantity to 2020 (%)	Historical cost (%)	Simulated cost (%)	Cost differential (%)
Austria	0.367	0.417	0.356	0.853
Belgium	1.366	1.306	1.378	1.056
Bulgaria	0.544	0.473	0.738	1.558
Cyprus	0.371	0.232	0.100	0.430
Czech Republic	2.070	3.364	3.093	0.919
Denmark	0.005	0.003	0.003	1.085
Estonia	0.001	0.001	0.001	0.620
Finland	0.001	0.001	0.002	1.631
France	7.091	6.668	5.336	0.800
Germany	49.633	59.926	55.863	0.932
Greece	3.467	2.405	2.925	1.216
Hungary	0.097	0.078	0.094	1.207
Ireland	0.016	0.019	0.018	0.957
Italy	11.572	7.904	8.603	1.088
Latvia	0.005	0.005	0.005	0.951
Lithuania	0.018	0.019	0.017	0.927
Luxembourg	0.101	0.044	0.053	1.207
Malta	0.052	0.037	0.048	1.293
The Netherlands	0.684	0.489	0.634	1.296
Poland	0.004	0.002	0.003	1.334
Portugal	1.769	1.010	1.318	1.306
Romania	0.384	0.269	0.307	1.141
Slovakia	0.360	0.375	0.342	0.912
Slovenia	0.167	0.093	0.117	1.266
Spain	17.168	12.232	16.061	1.313
Sweden	0.005	0.006	0.006	0.993
UK	2.686	2.621	2.578	0.984

particular, the function is used to simulate costs and allocation between EU countries of a fixed quantity target of RES. We take for example the target of photovoltaics to 2020 given by 2020 climate and energy package. Given the target to 2020 of 83390 GWh of photovoltaics production, we have simulated the optimal allocation among the EU 27 members, according to their relative price (LCOE) of production[4] (Table 5.4).

[4] We compute different LCOE for each technology in each country, since each country is characterized by a different level of solar radiation, wind intensity, and biomass productivity.

Notice in column 2, the actual, i.e., historical costs, and in column 3 the simulated costs resulting from the optimal allocation of photovoltaics production among the EU 27 members, which minimizes the total EU cost.

Our estimates show an important result: overall, this allocation would save almost 4 % of total EU cost. Of course, for some EU countries, there is a net saving and for others there is an increase in cost. For instance, Austria, Germany, Ireland, and the UK would pay less, while Belgium, Italy, and Spain would have to pay more. The reason is that countries with lower prices should produce more and vice versa. However, the overall EU cost is clearly lower.

The savings we get through our model, i.e., 4 % of total costs to reach 20 % RES target by 2020, is a significant number because it can positively influence the behavior of those who are involved in the politics of financial support to RES that are households. Indeed, acknowledging that RES goal imposes extra costs on households, several studies have investigated the relation between RES support policies and consumers' willingness to pay. Results show that a reduction of RES cost increases the share of favorable households, in the order of several percentage points (Bollino 2009; Scarpa and Willis 2010; Bigerna and Polinori 2014). Thus, we can infer that a reduction of the RES costs production can lead more likely to an increase in the political favor of households, i.e., those who ultimately vote consensus for RES development policies.

This result is really crucial in making possible to test the true spirit of cooperation among EU members. It is conceivable that there could be requests of compensations. Countries with higher costs would try to reduce their allocation but this will increase the cost for the whole EU. So they can try to negotiate some compensation.

Finally, it would be important in the future outlook to 2030 to take greater account of both infrastructural aspects and those related to research and development activities. Issues on which the Communication 2014 (0015) skates over. Instead, now it should be encouraged investments in those forms of flexibility (smart grid, demand management, storage tanks, and balancing services and flexibility), useful to integrate effectively RES into the system and in the electricity markets developed in the last decade in the EU. Even more, it is needed to focus European efforts to push in the direction of research and development activities for RES deployment.

It is only in this way that it will be possible for the EU to arrive at one or more technological breakthroughs (perhaps as part of the storage and smart grids) which, similarly to what happened with the extraction techniques in shale gas in the USA, are able to making RES an element of competitiveness in the energy markets.

5.5 Concluding Remarks

Climate change has become the main pillar of EU policies aimed at accelerating the transition towards sustainable development, a future with low-emission GHG. In this book, we propose an economic analysis of the current situation of RES with focus at the European level and prospects of RES development in the medium term in a sustainable development perspective.

We have provided an analysis on current EU policies, with targets up to 2050, and prior agreements that led to the current ones. The analysis starts from the recent agreement "Roadmap for moving to a competitive low-carbon economy in 2050" promoted by the EU, and focus on the 2030 framework for climate and energy policies presented on January 2014 by the European Commission (COM 2014(0015)) as an intermediate step to achieve the goal of longer term to 2050. The analysis has proceeded backwards, illustrating the main agreements that have led the EU to the current situation, that are the 2020 climate and energy package and the implementation of the Kyoto Protocol at EU level (Chap. 1).

We have provided the state of RES development in EU countries in terms of installed capacity and energy produced by each technology. The analysis carried out confirms that the production of energy from RES is a consolidated reality in the EU, able to meet 14 % of EU energy demand in 2013. Then we have compared the main world scenarios up to 2040 with a focus on the new EU policy scenarios (Chap. 2).

We have analyzed support mechanisms designed to encourage learning by doing and innovation in order to improve the competitiveness of RES compared to fossil fuels technologies. Drawing from experience in the EU energy system, we have offered a critical overview of existing policy supports for RES, with a focus to price-oriented mechanisms. We have evaluated the effect of RES support mechanisms in relation to the market structure (Chap. 3).

The objectives of mandatory RES share are implemented at the national level and they vary from one country to another; they are between 10 and 49 % of final consumption by 2020. We have offered an updated and comprehensive review in which each EU country stands in relation to 2020 RES' objectives (Directive 2009/28/EC). The EU has a huge potential to develop a new generation of energy technologies like offshore wind, solar energy, or technologies for the exploitation of biomass; we have illustrated possible RES technological breakthroughs which are currently discussed among scientists. Given the increased use of RES, we have provided a brief analysis on the smart grid technologies that are becoming a priority of the EU energy policy (Chap. 4).

Finally, we have proposed a new policy strategy which is based on a new RES development scenario. Our simulation model shows how to reach the RES quantity goal at minimum costs for the EU as a whole. Results show that with our mechanism costs incurred by the EU as a whole are lower than those implicated by the 2020 climate and energy package. Indeed, based on a logic of cost minimization for the EU, RES should be produced in countries where, for environmental vocations and technology, production costs are lower. This clearly involves an unequal concentration in the territories that have to be compensated through a fair mechanism. The decoupling of RES and cost share requires cooperation and compensation among EU countries. Our proposal might be a useful tool for the EU policy makers at the negotiation table among EU member countries for welfare improvement at EU level (Chap. 5).

Our policy implication strongly affirms that EU 28 countries should act as a single entity in such a way as to present the EU "as a single nation" in the world when sitting at the world negotiation tables. The EU in the course of the history has always been a leader in the intellectual, philosophical, and artistic fields with the Roman Empire, the Renaissance, and then with the industrial revolution.

Today, that leadership is in question in the environmental field since the EU is risking to become a residual area in the world. The EU currently represents only 11 % of global CO_2 emissions and in 2030 such share will be even lower given the rapid growth of emerging areas of the world. In the environmental field, the EU obviously cannot rely on a quantitative supremacy to impose its solution to the rest of the world. Nor the EU can exploit the economies of scale at the world level. The only way to be one of the major players on the world stage for the EU is to make continuous technological innovations to be successfully disseminated to the rest of the world. In this sense, the technological innovation should not be in "their own backyard" but it must be widespread and accepted as a new paradigm in the world. Otherwise "clean Europe" means that the EU produces less manufacturing in its territories and it imports dirtier products from the rest of the world with particular reference to the new emerging economies. In this way, the EU risks to lose competitiveness in its own manufacturing system while contributing to increase polluting emissions by buying abroad.

The real problem is that with all the restrictions, to which the EU must comply, it becomes more and more expensive to produce in the EU and then the EU buys goods produced where environmental standards are less stringent. There is here a dilemma. On the one hand, stringent environmental constraints impose higher and higher standards (costs) to European manufacturing system. On the other hand, these higher costs in Europe are also likely to strangle the best existing technologies that fail to keep up with these stringent environmental constraints.

Indeed, the EU risks to lose such technologies in favor of new locations that have less environmental constraints and therefore can afford to produce at lower costs. Leaving aside the classical issue of the manufacturing deindustrialization, let us focus on the case of the European refining sector. There is a danger of closing cutting-edge technologies in the EU because of the chocking environmental regulation. Consequently, the European demand for refined products will be fueled by neighboring non-European refineries characterized by lower environmental standards. This implies an increase in pollution at the world level because the goods' demand from Europe continues and a welfare loss for the European producers.

With these considerations in mind, we believe that the EU can remain one of the leading actors in the main of environmental policies by reunifying in a single European strategy the 28 fragmented national policies. To this end, let us not waste this opportunity by developing further the recent COM 2014 (0015) giving a clear mandate to the new European Commission that started the mandate at the end of 2014. In this sense, the EU might be able to overtake the USA still characterized by disharmonies between federal environmental policy and individual states policies.

The cornerstones of this new policy on the basis of our recommendations are as follows. First, the implementation of support mechanisms that are price driven instead of quantity driven in order to promote flexibility in the RES production choices. Second, a unique RES support scheme, shared at the European level, to spur RES localization taking into account the territorial vocation. Third, establishment of dynamic credibility of the policy framework over time.

With reference to this latter issue, we notice that at the firm level the technological innovation needs a stable incentives' path that supports new investments instead of chocking them. This can be done in only one way: policies have to be clear and maintained in the long term in such a way as to avoid the time-inconsistency problem (Kydland and Prescott 1977). We advocate a discontinuation of the policies of stop and go of the past, in favor of a certain long-term horizon of the RES incentive framework.

This highlights a crucial issue. Any level of incentives, maintained for a given period of time, will give an advantage to earn extra profits to those who manage to reach advanced technological solutions with lower costs compared to other competitors. Such extra profits are, on the one hand, a welfare loss for consumers, but on the other hand, the ultimate motivation to continue the technological innovation for firms who want to be first in the innovation race.

So the new EU policy strategies must be based on a coherent design which has to be both credible and consistent in the medium and long term in order to balance the trade-off between the defense of the consumer welfare and the spur of the firm profit.

At this end of this book, we would like to highlight that our analysis does not deliberately take into account many exogenous events that may influence RES development. The following issues should be kept in mind:

- The current international financial and economic crisis has involved all the EU countries with a negative impact in national economies. Its effects are still in place and the ongoing phenomenon deserves a wider reflection on future development scenarios. Although RES, in particular the wind and solar sectors, are now quite mature, governments across the EU are still burdened with the responsibility of supporting RES with funding since they are not yet able to compete with fossil fuels technologies. The economic crisis may change the perspective of the public opinion changing the priorities, lowering the attention to the climate change issue.
- Environmental policies often involve very long time horizons and it involves uncertainty about policies stability. Indeed, despite the best framework the policy maker can design, investments in RES might be suboptimal since investors do not perceive climate change policies as a long lasting government commitment. There have been occurred several turnabouts on climate change policies that support financially RES in a shaky way, as it has happened recently in USA, Germany, Italy, and Spain. Policy and regulatory uncertainties put RES at risk of falling quantitative and qualitative goals to meet global climate change objectives.

- Major technological developments, i.e., technological breakthroughs, are difficult to predict. This is so because their result from sudden adjustments to external influences or challenges to the single firm or to the firm context. For this reason, it is generally difficult to define the probability of realization and the time of occurrence of innovations. When major breakthroughs change the nature of technology fundamentally, this can lead to uncertain development paths (Brezis et al. 1993; Landes 1966): "Prosperity and success are their own worst enemies." This implies the risk that a technology that initially seems to bring a welfare improvement for the nation, can result in a miserable decline afterwards.

In conclusion, given economic and political uncertainties that influence RES diffusion at both EU and global levels, it is necessary to be aware and continuously monitor the process of the RES development. We believe that stability of policies is a fundamental value to continue to spur growth and prosperity, but at the same time, we acknowledge that this requires a continuous level of attention of policy refining and of market monitoring to avoid the risk of sudden change of course, sudden turns, and sudden failures in the growth of our way of life.

References

Barrett S (1994) Self-enforcing international environmental agreements. Oxf Econ Pap (Special Issue on Environmental Economics) 46:878–894

Bigerna S, Polinori P (2014) Italian households' willingness to pay for green electricity. Renew Sustain Energy Rev 34:110–121

Bollino CA (2009) The willingness to pay for renewable energy sources: the case of italy with socio-demographic determinants. Energy J 30(2):81–96

Bollino CA, Micheli S (2012) On the relative optimality of environmental policy instruments: an application of the work of Alberto Alesina. Atlan Econ J 40(4):385–399

Bollino CA, Micheli S (2014) Cooperation and free riding in EU environmental agreements. Int Adv Econ Res 20(2):247–248

Brezis ES, Krugman PR, Tsiddon D (1993) Leapfrogging in international competition: a theory of cycles in national technological leadership. Am Econ Rev 83(5):1211–1219

Carraro C, Eychmans J, Finus M (2006) Optimal transfers and participation decisions in international environmental agreements. Rev Int Organ 1(4):379–396

Enerdata (2014) Costs and benefits to EU member states of 2030 climate and energy targets. Final report from Enerdata's Global Energy Forecasting department completed for the UK's Department of Energy and Climate Change (DECC)

European Commission (2014) Communication (2014) 0015 from the Commission to the European Parliament, the Council, the European Economic and Social Committee and the Committee of the Regions. A policy framework for climate and energy in the period from 2020 to 2030

European Commission (2009) Directive 2009/28/EC of the European Parliament and of the Council of 23 April 2009 on the promotion of the use of energy from renewable sources and amending and subsequently repealing Directives 2001/77/EC and 2003/30/EC. Offl J Eur Union L 140:16–62

Eurostat (2014) http://epp.eurostat.ec.europa.eu/portal/page/portal/statistics/themes. Accessed 28 Jul 2014

Kydland FE, Prescott EC (1977) Rules rather than discretion: the inconsistency of optimal plans. J Polit Econ 85:473–492

Kolstad CD (2000) Environmental economics. Oxford University Press, Oxford
International Energy Agency (IEA) (2014a) World Energy Outlook 2014, OECD/IEA, Paris
International Energy Agency (IEA) (2014b) World Energy Investment Outlook 2014, OECD/IEA, Paris
Landes D (1996) Technological change and development in Western Europe, 1750–1914. In: Habakkuk H, Postan M (eds) The Cambridge economic history of Europe, vol VI. Cambridge University Press, Cambridge, pp 274–585
Nordhaus WD (2009) The impact of treaty nonparticipation on the costs of slowing global warming. Energy J 30(2):39–52
Rathman M (2007) Do support systems for RES-E reduce EU-ETS-driven electricity prices? Energy Policy 35 (1):342–349
Scarpa R, Willis K (2010) Willingness-to-pay for renewable energy: primary and discretionary choice of British households' for micro-generation technologies. Energy Econ 32(1):129–136
Stavins RN (1995) Transaction costs and tradeable permits. J Environ Econ Manag 29(2):133–148

Index

S. Bigerna et al., *The Sustainability of Renewable Energy in Europe*,
DOI 10.1007/978-3-319-12343-1

MIX
Papier aus verantwortungsvollen Quellen
Paper from responsible sources
FSC® C105338

If you have any concerns about our products,
you can contact us on
ProductSafety@springernature.com

In case Publisher is established outside the EU,
the EU authorized representative is:
Springer Nature Customer Service Center GmbH
Europaplatz 3, 69115 Heidelberg, Germany

Printed by Libri Plureos GmbH
in Hamburg, Germany